BUSINESS ISSUES, COMPETITION AND ENTREPRENEURSHIP

INNOVATORS AND TINKERERS ROLE IN BUSINESS: A GENERAL AND TECHNICAL ANALYSIS

BUSINESS ISSUES, COMPETITION AND ENTREPRENEURSHIP

Improving Internet Access to Help Small Business Compete in a Global Economy
Hermann E. Walker (Editor)
2009. ISBN: 978-1-60692-515-7

Multinational Companies: Outsourcing, Conduct, and Taxes
Loran K. Cornejo (Editor)
2009. ISBN: 978-1-60741-260-1

Private Equity and its Impact
Spencer J. Fritz (Editor)
2009. ISBN: 978-1-60692-682-6

Progress in Management Engineering
Lucas P. Gragg and Jan M. Cassell (Editor)
2009. ISBN: 978-1-60741-310-3

Codes of Good Governance around the World
Felix J. Lopez Iturriaga (Editors)
2009. ISBN: 978-1-60741-141-3
2010. ISBN: 978-1-60876-785-4
(E-book)

Crisis Decision Making
Chien-Ta Bruce Ho, KB Oh, Richard J. Pech, Geoffrey Durden and Bret Slade
2009. ISBN: 978-1-60876-073-2

Electronic Breadcrumbs: Issues in Tracking Consumers
Dmitar N. Kovac
2010. ISBN: 978-1-60741-600-5

Handbook of Business and Finance: Multinational Companies, Venture Capital and Non-Profit Organizations
Matthaus Bergmann and Timotheus Faust (Editors)
2010. ISBN: 978-1-60692-855-4

Human Behavior and the Corporate Social Responsibility of Firm Leaders
Mikko H. Manner
2010. ISBN: 978-1-60876-834-9

The U.S. Auto Industry and the Role of Federal Assistance
James R. Elliot (Editor)
2010. ISBN: 978-1-60741-322-6

Global Operations Management
Lee R. Hockley
2010. ISBN: 978-1-60876-355-9
2010. ISBN: 978-1-61668-351-1
(E-book)

Small Business Tax Issues and Benefits
Alton E. Kastenberg (Editor)
2010. ISBN: 978-1-61668-806-6
2010. ISBN: 978-1-61668-393-1
(E-book)

Supply-Chain Management: Theories, Activities/Functions and Problems
Regina M. Samson (Editor)
2010. ISBN: 978-1-61668-284-2
2010. ISBN: 978-1-61668-715-1
(E-book)

Group Processes in Ethnically Diverse Organizations: Language and Intercultural Learning *
Jakob Lauring and Charlotte Jonasson
2010. ISBN: 978-1-61668-761-8
2010. ISBN: 978-1-61728-325-3
(E-book)

Competitive Advantage in the Contracting Business
Low Sui Pheng, Faisal Manzoor Arain and Lim Ting Ting
2010. ISBN: 978-1-60876-800-4

Innovators and Tinkerers Role in Business: A General and Technical Analysis
Steven C. Hymes (Editor)
2010. ISBN: 978-1-60876-815-8

Immigrant Entrepreneurship in the U.S.
Christian D. Knowles (Editor
2010. ISBN: 978-1-60876-816-5

The Small Business Economy
Sally N. Hayes (Editor)
2010. ISBN: 978-1-60741-468-1
2010. ISBN: 978-1-61668-654-3
(E-book)

BUSINESS ISSUES, COMPETITION AND ENTREPRENEURSHIP

INNOVATORS AND TINKERERS ROLE IN BUSINESS: A GENERAL AND TECHNICAL ANALYSIS

STEVEN C. HYMES
EDITOR

Nova Science Publishers, Inc.
New York

Copyright © 2010 by Nova Science Publishers, Inc.

All rights reserved. No part of this book may be reproduced, stored in a retrieval system or transmitted in any form or by any means: electronic, electrostatic, magnetic, tape, mechanical photocopying, recording or otherwise without the written permission of the Publisher.

For permission to use material from this book please contact us:
Telephone 631-231-7269; Fax 631-231-8175
Web Site: http://www.novapublishers.com

NOTICE TO THE READER

The Publisher has taken reasonable care in the preparation of this book, but makes no expressed or implied warranty of any kind and assumes no responsibility for any errors or omissions. No liability is assumed for incidental or consequential damages in connection with or arising out of information contained in this book. The Publisher shall not be liable for any special, consequential, or exemplary damages resulting, in whole or in part, from the readers' use of, or reliance upon, this material. Any parts of this book based on government reports are so indicated and copyright is claimed for those parts to the extent applicable to compilations of such works.

Independent verification should be sought for any data, advice or recommendations contained in this book. In addition, no responsibility is assumed by the publisher for any injury and/or damage to persons or property arising from any methods, products, instructions, ideas or otherwise contained in this publication.

This publication is designed to provide accurate and authoritative information with regard to the subject matter covered herein. It is sold with the clear understanding that the Publisher is not engaged in rendering legal or any other professional services. If legal or any other expert assistance is required, the services of a competent person should be sought. FROM A DECLARATION OF PARTICIPANTS JOINTLY ADOPTED BY A COMMITTEE OF THE AMERICAN BAR ASSOCIATION AND A COMMITTEE OF PUBLISHERS.

LIBRARY OF CONGRESS CATALOGING-IN-PUBLICATION DATA

Innovators and tinkerers role in business : a general and technical analysis / editor, Steven C. Hymes.
    p. cm.
 Includes bibliographical references and index.
 ISBN 978-1-60876-815-8 (softcover : alk. paper)
 1. Technological innovations--Economic aspects. 2. Diffusion of innovations--Economic aspects. 3. Small business--Valuation. 4. Patents. 5. Inventions. I. Hymes, Steven C.
 HC79.T4I56535 2009
 338'.064--dc22
                                                                2009046468

Published by Nova Science Publishers, Inc. ✦ *New York*

CONTENTS

Preface		ix
Chapter 1	Innovation in Small Businesses: Drivers of Change and Value Use *C.J. Isom and David R. Jarczyk*	1
Chapter 2	Network of Tinkerers: A Model of Open-Source Technology Innovation *Peter B. Meyer*	31
Chapter Sources		61
Index		63

PREFACE

This book investigates various drivers of innovation within small businesses, as well as the role that innovation plays in creating value in small businesses. The analysis suggests that additions in employee headcount increase innovation while growth in sales does not increase innovation. The analysis also finds that increases in research and development ("R&D") expenditures enhance small business value in certain industries, but not uniformly and not in all the industries investigated. The authors of this book find that the number of patents owned by a small business is not a good indicator of a firm's value. Moreover, this book has a model of agents called tinkerers who want to improve a technology for their own reasons, by their own criteria, and who see no way to profit from it. Under these conditions, they would rather share their technology than work alone. This book places open source technology development in an abstract, deductive model. Within the model, a new industry appears when a tinkerer envisions a way to profit from the technology, and leaves the network to try that. In the cases described, amateur tinkerering eventually led to increases in commercial output and productivity. This book consists of public documents which have been located, gathered, combined, reformatted, and enhanced with a subject index, selectively edited and bound to provide easy access.

Chapter 1 - This paper investigates various drivers of innovation within small businesses, as well as the role that innovation plays in creating value in small businesses. The analysis suggests that additions in employee headcount increase innovation while growth in sales does not increase innovation. The analysis also finds that increases in research and development ("R&D") expenditures enhance small business value in certain industries, but not uniformly and not in all the industries investigated. Finally, the paper finds that the number of patents owned by a small business is not a good indicator of a firm's value.

Chapter 2 - Airplanes were invented by hobbyists and experimenters, and some personal computers were as well. Similarly, many open-source software developers are interested in the software they make, and not focused on profit. Based on these cases, this paper has a model of agents called tinkerers who want to improve a technology for their own reasons, by their own criteria, and who see no way to profit from it. Under these conditions, they would rather share their technology than work alone. The members of the agreement form an information network. The network's members optimally specialize based on their opportunities in particular aspects of the technology or in expanding or managing the network. Endogenously there are incentives to standardize on designs and descriptions of the technology. A tinkerer in the network who sees an opportunity to produce a profitable product may exit the network to create a startup firm and conduct focused research and development. Thus a new industry can arise.

In: Innovators and Tinkerers Role in Business
Editors: Steven C. Hymes

ISBN: 978-1-60876-815-8
© 2010 Nova Science Publishers, Inc.

Chapter 1

INNOVATION IN SMALL BUSINESSES: DRIVERS OF CHANGE AND VALUE USE

C.J. Isom and David R. Jarczyk

ABSTRACT

This paper investigates various drivers of innovation within small businesses, as well as the role that innovation plays in creating value in small businesses. The analysis suggests that additions in employee headcount increase innovation while growth in sales does not increase innovation. The analysis also finds that increases in research and development ("R&D") **expenditures** enhance small business value in certain industries, but not uniformly and not in all the industries investigated. Finally, the paper finds that the number of patents owned by a small business is not a good indicator of a firm's value.

EXECUTIVE SUMMARY

Innovation and firm value are key drivers for business success. Their respective roles are vital in creating and improving goods and services, developing market demand, meeting market expectations, and increasing shareholders' **wealth**.

In this report we examined the drivers of innovation within small businesses. We examine what drivers affect the number of patents issued to a small business by using patent production as a proxy for innovation; these drivers include employee headcount, sales, R&D expenditures, and other factors. We also examined the factors that affect firm value: R&D expenditures, patent issuance, and others.

Our first result shows that innovation increases significantly as small businesses increase employee headcount. Using a firm's patent activity as a proxy for innovative activity, our empirical results show that a one-hundred person increase in employee headcount increases innovative activity by 20 percent. Perline, Axtell, and Teitelbaum (2006) found similar results with respect to the survival rates of firms and the firm size. Wallsten (2000) also found that firms with more employees win more Small Business Innovation Research grants. Our analysis is consistent over every year for which research was performed (2004, 2005 and 2006), using a multivariate regression analysis dating back to 2000.

In our second result, we observe that changes in a firm's sales have neither a positive nor negative effect on innovation. The analysis, as before, applies to 2004 through 2006, and uses patent activity as a proxy for innovative activity.

Thirdly, we observe that empirical results reveal that there is no statistical relationship between patent count and market value. Lerner (1994) found different results when he modeled the tradeoff between patent protection and a firm's valuation. While Lerner's study focused on patents within the biotechnology industry, our analysis uses an industry-by-industry approach, focusing on five specific industries:[1]

- Chemicals and Allied Products ("Chemicals");
- Industrial and Commercial Machinery and Computer Equipment ("Industrial Machinery");
- Electronic and Other Electrical Equipment and Components, Except Computer Equipment ("Electronics");
- Measuring, Analyzing, and Controlling Instruments; Photographic, Medical, and Optical Goods; Watches and Clocks ("Mechanical Goods"); and
- Business Services ("Business Services").

Perhaps one explanation for this result is that patent count, per se, may be necessary to protect a position in an industry versus creating ground-breaking opportunities for a firm. Alternatively, patents may not create great leaps of technological capabilities, or some patents may have limited application. Therefore the market may not reward patent generation per se, even though patent generation is an indicator of innovation.

Finally, our results indicate that there may be as much as a three percent increase in market value for every ten percent increase in R&D expenditures. However, this relationship is dependent upon industry.

The remainder of this paper explains some of our analysis in greater detail, and concludes with areas of further research that may be warranted.

1. INTRODUCTION

Much research has been performed on the role of small businesses in the US economy. These topics include small businesses' impact on innovation (Schumpeter 1934), the effects of government grants awarded to small businesses (Wallsten 2000), and innovation impact on a small business' value (Lerner 1994). Although there is an abundance of research, the availability of data limits the extent of research that can be performed. This limit in available resources and data has led to creative studies, but with the emergence of more information researchers are able to expand their understanding.

Innovation has been a constant proxy for measuring small business success. Patent production has been the most common proxy to measure innovation because data is readily accessible. Cincera (1997) concluded that R&D expenditures and technological spillovers were the result of patent production for all firms. While his research found determinants of patent production for all firms, it does not address how the size of the firm can impact patent production. Acs and Audretsch's (1988) results show that determinants of innovation are positively related to R&D expenditures and skilled labor. However, their research did show that these determinants have disparate effects on small firms.

Along with innovation, a significant body of research has looked at valuation of small businesses. Lerner's (1994) study on biotechnology firms found that the breadth of patent protection positively affects a firm's valuation. While this research provides results on the relationship between patent protection and firm value, it does not address the relationship between innovation and firm value of small businesses in different industries.

Using data from *ktMINE*, Standard & Poor's *Compustat* database, and the U.S. Patent and Trademark Office ("USPTO") databases, Ceteris constructed a dataset containing nearly 22,000 observations with detailed information from 2000 to 2006 on the size of small businesses, the industries in which they compete, the intensity of their patenting activities, and their financial performance.

While other research has addressed topics such as innovation, patent production, R&D expenditures, and firm value, no other report has performed analysis solely on small businesses using the data that we obtained.

The first thing we examined is the relationship between employee headcount and innovation within a small business. We measured innovation by patent production. This analysis is performed on an industry-wide basis and limited to small firms. Observing these firms in the years 2004, 2005, and 2006, our findings show that there is a positive relationship between employee headcount and patent production in all of the years observed.

We then explored the relationship between sales and innovation within a small business. Like the previous study, we use patent production as a proxy for innovation. The analysis shows that there is no significant relationship between sales and patent production. The years observed for this study were 2004, 2005 and 2006.

Our next analysis focused on the relationship between patent production and firm value. While Griliches' (1990) work suggests that efforts to explain the level of stock market valuations using patent measures have been disappointing, this paper is focused on the effects of patent production on the value of a small business within certain observed industries. Using data from **five industries and measuring firm value as the firm's price**-to-book ratio, we find that there is no significant relationship between patent production and firm value.

Finally, we examined the relationship between R&D expenditures and firm value. Continuing to use the price-to-book ratio as a proxy for firm value, we discovered that in three of the five observed industries, R&D expenditures have a positive effect on firm value.

2. LITERATURE REVIEW

Small firms play an instrumental role in the U.S. economy, employing half of all private sector employees and creating more than half of non-farm private Gross Domestic Product.[2] Further, small businesses that innovate are most likely to grow into large businesses and become a source of highly technical, high-paying jobs in the future. As such, understanding the trends of small businesses, especially patenting trends, is becoming a focal point in applied economics and policymaking. However, detailed research of this type is not readily available.

Much research has been done in the field of economics measuring patents as indicators of innovation and technological change. From a policymaking perspective, the literature focuses on how to design an optimal patent policy. From an economic perspective, studies examine the importance of innovation in the economy, the role of small firms in the innovation process, and the role of government in promoting innovation.

There are also research papers that investigate the relationship between firms' **patent portfolios** and their financial performance. However, these studies are often based on samples of large businesses. Small firms tend to be overlooked, because it is much more difficult to obtain detailed and reliable financial information. The difficulty lies in the fact that the majority of small firms are privately-owned. Such firms are not required to file their financial statements with the U.S. Securities Exchange Commission ("SEC"), have their statements audited by an accounting firm, or offer information publicly.

However, smaller firms play an important role in technological change. Smaller firms have many advantages as sources of innovation because they are quick to adopt new and high risk initiatives; they facilitate structures that value ideas and originality; and they have a better capacity to reap substantial rewards from market share in small niche markets. Statistically, small firms outperform large firms in terms of patents per employee by 13 to 14 times.[3]

Theories on the role of small firms in the economy have been developed since the early 1900's. Schumpeter (1934) considered small firms an important venue for innovation and change to be incorporated into the economic system. Furthering this idea, scholars have attempted to link the size of firms with their innovation capability. However, empirical studies addressing these theories have been somewhat contradictory. While some studies have found a positive relationship between firm size and innovative capability, others have identified no relationship-or even a negative one (Audretsch and Acs 1991). The most extensive studies found a U-shaped relationship between firm size and innovative intensity in terms of innovations per employee (Audretsch and Acs 1991). That is, innovative intensity was strongly affected by small and large firms; however, in middle-size firms, the impact was weaker. This result is not robust because the results change when non-innovative firms are included in the sample. Perline, Axtell and Teitelbaum (2006) found that over longer time frames, there are fewer large growth changes within a business than observed in shorter time frames. Moreover, employment swings within a business are more likely to occur in shrinking firms than in those that are expanding.

Studies on R&D expenditures and their impact on innovation have also been analyzed. Wallsten (2000), using a multi-equation model, determined that firms with more employees who appear to do more research win more Small Business Innovation Research grants. Klette and Griliches (2000) observed that R&D expenditures are proportional to the firm's sales. Cincera (1997) suggested a positive impact on the technological spillovers on a firm's own innovation.

None of the aforementioned papers or literature, to our knowledge, provides a detailed focus on the effect that innovation has on market value, or the factors that drive innovation within firms. The approach that we take in this paper provides new analysis for understanding certain drivers and outcomes of innovative activity among small businesses, which comprise a large part of the U.S. economy.

3. DATA

3.1. Introduction

This section details the data gathering activities to construct the dataset for our analysis. Information was gathered from three primary sources:

- Standard & Poor's *Compustat* database;
- the U.S. Patent and Trademark Office's databases; and
- *ktMINE*.

3.2. Standard & Poor's *Compustat* Database Review

Ceteris examined the Standard & Poor's Compustat database (September 30, 2007) for public companies that are traded on U.S. stock exchanges and Over-the-Counter markets. This database provides business descriptions, financial data, and other company-specific data for these companies. Databases produced by Standard & Poor's are commonly used to analyze companies by a variety of financial industries.

In order to limit our dataset to small business results, the Standard & Poor's Compustat database was screened for companies that had less than 500 employees for fiscal year 2006. In order to identify the search for companies with less than 500 employees, the database was screened using the employee mnemonic[4] "EMP," as defined by Standard & Poor's Compustat. This screening led to the identification of 3,659 companies classified as small businesses.

Next, we used the Standard & Poor's Compustat database to acquire the following list of financial information for the 3,659 companies:

- Employees;
- Sales – Net;
- R&D Expenditures; and
- Price to Book – Fiscal Year End.

For a complete listing of terms along with their mnemonics used to extract data for this analysis, see Appendix 2. Appendix 2 also includes the definitions and/or calculation used by Standard & Poor's Compustat for each financial item.

Lastly, we performed a final review on the employee data. For fiscal years 2005 and 2006, each company needed to have less than or equal to 500 employees. Furthermore, Compustat lists data as "@NA" if no information is available for the company for a given year. If a company had a "@NA" in the employee field, it was screened out of the total company list. This employee screen removed an additional 612 companies. As a result, we identified 3,047 companies for use in the next phase of acquiring patent counts.

3.3. Patent Count Review

Ceteris relied on databases provided by the USPTO to obtain patent count information for the 3,047 small businesses that we selected earlier. The "Advanced Search" function within the USPTO search engine allows the database to be searched by individual fields that describe each patent. These fields include Application Date, Application City, Issue Date, Patent Number, Assignee Name and more. Searches under Assignee Name and Issue Date allowed Ceteris to identify the total patents owned by the specified companies and patents issued to those companies within calendar years 2004, 2005 and 2006.[5]

Finally, after gathering the patent information, we filtered the data to keep companies that had data in each of the aforementioned categories from the S&P database dating back to 2000. This screening process removed 2,410 companies, resulting in 637 companies to form the sample of interest for this study. For each of the remaining companies, detailed company and financial information was at our disposal, including number of employees, sales, R&D expenditures and price to book value.

3.4. ktMINE

ktMINE is an online tool that provides detailed information related to intellectual property. This information includes the specific detail of licensing transactions as reported by U.S. public companies and information related to the underlying intangibles, including patents, exploited through these agreements. We examined *ktMINE* to identify small businesses that license patented technologies. Then we compared the identified companies to the companies derived from the previous steps to ensure no companies were missing from the dataset. The completion of this step allowed us to be confident that the dataset compiled was accurate and complete.

4. RESEARCH METHODOLOGY

4.1. Introduction

As mentioned at the outset of this paper, our research sought to answer specific questions related to relationships between specific potential drivers of innovation – namely, employee headcount and sales, as well as the effect of different measurements of innovation (patent count and R&D expenditures) on market value. We addressed each of these questions by identifying specific forms of the relationships between the variables, and ultimately we strove to identify the best "fit" (i.e., the model with the strongest relationship to the observed data that was also theoretically sound) to address each question.

4.2. The Relationship between Employee Headcount and Sales and Innovation

While we address the relationship between employee headcount and innovation and the relationship between sales and innovation separately, our evaluations of the relationships flow from the same econometric model. This allowed us to control for any interaction or combined effects that these two variables might have on innovative activity.

In investigating the relationship between employee headcount and innovation, we posited that additions of employees in one year may not necessarily contribute to increases in innovation immediately, particularly when using patent output as a proxy for innovative activity. Other studies (Acs and Audretsch, 1988; Jaffe, 1986; Scherer, 1982; and Mansfield, 1981) have concluded that innovation output is related to innovation-inducing inputs in the previous period. We followed a similar approach, but modified it to fit our model. First, for an employee to add to the innovative process, it may take time for the employee to understand the research agenda of, and challenges faced by, the firm in which they are employed; in other words, an employee may need to move up the learning curve before adding to the innovative activity of the firm. Second, there is a lag between initial innovation and the time at which a patent is awarded.

Therefore, to account for this effect, we created a lagged model to measure the effect that employment might have on patent production and observed that innovative activity can be enhanced up to five years after employee headcount increases.

In a general regression model, marginal effects of the independent variable are one-time events. The response to the dependent variable is immediate and is completed at the end of the measuring period. For dynamic models, the marginal effects of the independent variable in any time period will affect the dependent variable in a future time period. The effect is not a one-time event, and it is not necessarily immediately observed.

A general form of a dynamic regression model can be defined as:

$$y_t = \alpha + \beta x_{t-1} + \varepsilon_t. \tag{1}$$

This model shows that a change in y is dependent on the changes in x from the previous period. By recognizing that past changes in the independent variable affect the dependent variable in this period, this function provides a better fit for the analysis.

We posited that there also might be a lag between sales and innovative activity. Increases in sales might create additional opportunities for firms, particularly smaller firms, where increases in sales may make it easier to obtain financing to fund future innovation.[6]

After determining the optimal number of lagged years to apply to our model, we used a modified moving average model applied to an ordinary least squares regression. By correcting the econometric form of the model, we were able to capture the lag effects of the coefficients while eliminating heteroscedasticity and multicollinearity.

To understand the rationale, let us observe equation (1) with more lagged variables. The model is presented as:

$$y_t = \alpha + \beta_i x_{t-1} + \beta_i x_{t-2} + \beta_i x_{t-3} + \phi_i z_{t-1} + \phi_i z_{t-2} + \phi_i z_{t-3} + \varepsilon_t. \quad (2)$$

Given this model, we see that the dependent variable, y_t is affected by changes in x and z in time periods $t-1$, $t-2$, $t-3$, and $t-4$. With the number of lags and cross-sectional data, there is concern of heteroscedasticity.[7] A moving average model is used as an alternative to the usual time-series process.[8]

Using the error term from equation (2), we can write the functional form of the error term as:

$$\varepsilon_t = v\varepsilon_{t-1} + u_t, \quad (3)$$

where

$$E[u_t] = 0$$
$$E[u_t^2] = \sigma_u^2$$

In the moving-average form, "each disturbance embodies the entire past history of the u's (Greene 258)". This embodiment of the past history eliminates the heteroscedasticity as well as autocorrelation.

While analyzing the effects that employee headcount and sales have on innovation, we posited that there also might be other factors that are affecting innovative activity. To eliminate any over estimation of the employee headcount and sales statistic, we included two variables that would provide the best fit for the estimation.

The first variable that we added to the regression was R&D expenditures. Previous studies have shown that there is a direct correlation with R&D expenditures and innovation (Bound, Cummins, Griliches, Hall and Jaffee 1984). We find it necessary to include R&D expenditures in the regression, given our use of patent count as a proxy for innovation.

The second variable we included is the ratio of price to book value. The price to book value ratio is a measurement of a firm's market value. We saw it appropriate to control for changes in innovation while measuring the market value of the firm. Because the ratio of price to book value measures the market's valuation of a firm, it provides information on what the market believes each firm is worth as well as the premium the market is willing to pay for the embedded intangible property, including efficiencies, within the firm. Efficient behavior directly impacts a company's value.[9]

The moving-average form is one that embodies the entire past history of the disturbances of the independent variables with the most recent observations, with greater weights given to recent observations versus the distant past (Greene 2003). The form we used for this study differs in that an equal weight is placed on the disturbances. By allowing for this distribution across the disturbances, we can use the average of the independent variable to provide consistent and efficient estimates. This is based on the assumption that past shocks will not vary across time.[10] For example, the effect that R&D expenditures in year $t-3$ has on patent production in year t is not necessarily less than the effect that R&D expenditures in year $t-2$. In fact, one could argue that the effect is greater in year $t-3$ than in year $t-2$. The alternate could be argued for employee headcount. Because of the ambiguity of past shock effects, we applied equal weights across all shocks, which will not affect the consistency of the regression. After taking the average of the variables, we apply an ordinary least squares regression to produce consistent estimates.

The model used in the analysis is as follows:

$$Patents_{i,t} = \beta_0 + \beta_1 \overline{employee}_{i,n} + \beta_2 \overline{sales}_{i,n} + \beta_3 \overline{research}_{i,n} + \beta_4 \overline{book}_{i,n} + \varepsilon_{i,n} \quad (4)$$

Where,

- *Patents$_t$* is the number of patents issued in the United States for firm i in year t;

- $\overline{employee}$ is the five-year average of employee headcount for firm i for the period n;
- $\overline{sales}$ is the five year average of net total sales in firm i for the period n;
- $\overline{research}$ is the five year average of R&D expenditures in firm i for the period n;
- $\overline{book}$ is the five year average of price to book values in firm i for period n;
- β_0 is the intercept and β_1, β_2, β_3, and β_4 are the coefficient estimates; and
- $\varepsilon_{i,n}$ is the error term.

Sections 5.2 and 5.3 explain the results as applied to this model.

4.3. The Relationship between Patent Production and Market Value

Our third analysis focuses on the relationship between innovation and firm market value, using patent production as a proxy for innovation and the price to book value ratio of firms as the measure of market value. As previously stated, measurement of market value provides signals of the embedded intangible property, including efficiencies, of the firm. The efficiency is **representative of the firm's management** and production capabilities. The price to book value signals the capabilities of the firm.

We posited that increases in patent production might have a positive effect on market value. Our evaluation of the data led to our belief that the dependent variable, the price to book value ratio, fits a non-linear function. Specifically, we applied a semi-log regression to measure the elasticity of the price to book value with changes in patent production. The primary result for the semi-log regression estimate is that it retains its consistency, efficiency, and asymptotic normality.

Because we used a non-linear model, we assumed that there was an underlying probability distribution for the observable y_i and a true parameter vector, B. We use the functional form:

$$\frac{y_i^\lambda - 1}{\lambda} = \alpha + \beta x_i + \varepsilon_i \tag{5}$$

Where y_i is the dependent variable, α is a constant, β is the parameter estimate, and λ is a given value. By applying this form to the study, the model corrects for heteroscedasticity and autocorrelation.[11]

For our model, the ratio of patents issued in year t to the total patent count per company is regressed against the price to book ratio. This ratio of patent production provides the percentage change of patent production for a given firm in a given year.[12]

The specific model[13] applied is:

$$\ln book_{i,j,t} = \gamma_0 + \gamma_1 pratio_{i,j,t} + \varepsilon_{i,j,t} \tag{6}$$

Where

- $\ln book_{i,j,t}$ is the natural logarithm of the price to book value for firm i, in industry j, for year t;
- $pratio_{i,j,t}$ is the ratio of patents issued in year t, divided by total patents for firm i, in industry j; and
- γ_0 is the intercept and γ_1 is the coefficient estimate for this model.

Section 5.4 presents the results as applied to this model.

4.4. The Relationship between R&D Expenditures and Market Value by Industry

For the final application, we used R&D expenditures as a proxy for innovative activity to determine how market value might be affected. We used the price to book value as a proxy for the intangible value of a firm, as the book value of a firm captures only the value of assets on the balance sheet of a firm, and internally created intangibles would not appear as an asset under generally accepted accounting principles. Therefore, we would expect R&D expenditures to increase the price to book value of a firm, provided the R&D activities were successful in creating intangibles, such as patents, trade secrets, enhanced production processes, etc., that investors would expect to increase a firm's market value in the marketplace. In our regression, we use the ratio of R&D expenditures to total net sales to correct for the relative scale of innovative activity. It also allowed us to eliminate outliers in the data.

Finally, we applied our analysis by industry, as different industries would be expected to have different expected levels of R&D expenditures. This also allows us to compare the relative benefits of R&D expenditures across industries.

We applied a log-linear model for this relationship, which provides consistent and efficient estimates.[14]

$$\ln book_{i,t,j} = \varphi_0 + \varphi_1 \ln r\_d\_sales_{i,j,t} + \varepsilon_{i,j,t}, \tag{7}$$

Where:

- $ln\ book_{i,j,t}$ is the natural logarithm of the price to book value for firm i, in industry j, for year t;
- $r\_d\_sales_{i,j,t}$ is the natural logarithmic value of the percentage change in R&D expenditures over net sales for firm i in industry j in year t; and
- φ_0 is the intercept and φ_1 is the coefficient estimate for this model.

Section 5.5 presents the results as applied to this model.

5. EMPIRICAL RESULTS

5.1. Introduction

This section presents the empirical results of the regression analyses presented in the prior section. Full details of our results are provided in the tables and appendices at the end of this report.

5.2. The Relationship between Employee Headcount and Innovation

Coefficient estimates of the independent variables are contained in Table 1, along with the t-statistic. The model controls for employee headcount, sales, R&D expenditures, and the price to book value ratio.[15]

Table 1 lists the results. As noted previously, employee headcount, which is based on a five-year average, increases innovative activity in a significant manner. We have evaluated this every year from 2004 to 2006 and have noted a positive relationship in each regression. Moreover, the relationship grew stronger over time.

In 2005 and 2006, increasing average employee headcount by one individual within a small business increases patent production by 0.002. It is important to note that, while this is a small value in itself, it is the signicance of the value that matters. First, one must remember that there is great value in the production of one patent, and one also would assume that there also is additional non-patentable but innovative activity that likely goes with the production of a patent.

Table 1. Employee Headcount and Sales Coefficient Estimates

Variable	2006		2005		2004	
	Estimate	T-Statistic	Estimate	T-Statistic	Estimate	T-Statistic
Employee Headcount	0.0028	2.38	0.00245	2.36	0.0011	1.21
Sales	0.00162	0.86	0.015	0.81	0.00283	1.18

A study done by Breitzman and Hicks (2008) shows that small firms develop more patents **per employee than larger ones**. Breitzman and Hick's results differ from our results because their analysis compares smaller firms to large firms.

Other results such as Perline, Axtell and Teitelbaum (2006) and Wallsten (2000) show that increases in employee headcount have a positive effect on innovation.

Our results differ in that we focused only on small firms. Within the sample of small firms, the more employees a firm has, the more patents will be produced.

5.3. The Relationship between Sales and Innovation

Our analysis of the relationship between sales and innovation provided very different results from the preceding section. Specifically, one finds that there is an insignificant effect between sales and patent production, as shown in Table 1.

The model we used for this estimate measured for the lagged effects that sales may have on innovation. However, it did not produce significant results.

One possible reason for this effect is that there is a high degree of variation in sales across firms relative to patent production. While we investigated many different forms of sales relative to patent production, we were unable to find a meaningful correlation within our dataset. If the conclusion that sales do not have a predictable effect on innovative activity is correct, this might imply that innovative activity would be greater in economies pursuing strategies of low unemployment versus high sales growth, *ceteris paribus*. Such a conclusion, however, would require additional research beyond the scope of this paper.

5.4. The Relationship between Patent Count and Market Value

The results of our analysis of patent production on market value for each observed industry are found in Table 2. The results show that patent issuance does not have any significant statistical effect on market value, as measured by the price to book ratio, within any industry.

From a statistical standpoint, these results may imply that market value is driven by factors other than patent count. Particularly because this analysis controls for variations across industries, the results of this conclusion may be that patents might not necessarily create enhanced value for firms. For example, a firm might create patents in order to maintain a competitive position or to block future long-term competition. This type of activity might not generate incremental earnings for a firm but would simply maintain the firm's status quo in an industry. Therefore, the market would not likely pay a premium for such activity.

An earlier study by Lerner (1994) found that there was an increase in a firm's value with an increase in patent scope. While these results examined the breadth of patent protection, our results focused solely on the production of patents.

Table 2. Patent Production Coefficient Estimates for Each Industry

Industry	2006 Estimate*	2006 T-Statistic	2005 Estimate*	2005 T-Statistic	2004 Estimate*	2004 T-Statistic
Chemicals and Allied Products	-1.73 (0.6564)	-0.26	0.2084 (0.8749)	0.24	0.558 (0.4944)	1.13
Industrial and Commercial Machinery and Computer Equipment	0.6564 (0.7819)	-0.83	-1.254 (1.384)	-0.91	-0.331 (0.75)	-0.44
Electronic and Other Electrical Equipment and Components, Except Computer Equipment	-0.7556 (0.5845)	-1.29	1.073 (0.736)	1.46	-0.4422 (0.7819)	-0.57
Measuring, Analyzing, and Controlling Instruments: Photographic Medical and Optical Goods; Watches and Clocks	-0.4932 (0.6884)	-0.72	2.909 (0.8847)	0.33	0.2909 (0.8847)	0.33
Business Services	0.06263 (4.5930)	0.14	-0.7551 (0.8232)	-0.92	0.0978 (0.6527)	0.15

* Standard errors in parenthesis

5.5. The Relationship between R&D Expenditures and Market Value by Industry

The results for our analysis presented in Table 3 show that for the years considered, R&D expenditures in the Industrial Machinery industry had the largest effect on firms' values. Firms in this industry engage in the manufacture of engines and turbines; farm and garden machinery; construction; mining; and oil field machinery; elevators and conveying equipment; industrial trucks and tractors; computer and peripheral equipment; and office machinery. In 2006, for every one percent increase in the percentage change in R&D expenditures to net sales, the price to book value increased by 0.3 percent for a given firm in this industry.

Table 3. R&D Expenditures Coefficient Estimates for Each Industry

Research and Development	2006 Estimate*	2006 T-Statistic	2005 Estimate*	2005 T-Statistic	2004 Estimate*	2004 T-Statistic
Chemicals and Allied Products	0.0631 (0.3780)	1.67	0.0479 (0.0348)	1.38	0.5287 (0.0318)	1.66
Industrial and Commercial Machinery and Computer Equipment	0.3143 (0.0741)	4.24	0.2062 (0.0733)	2.81	0.1826 (0.0570)	3.21
Electronic and Other Electrical Equipment and Components, Except Computer Equipment	0.1224 (0.0436)	2.81	0.1534 (0.0588)	2.61	0.1217 (0.0606)	2.01
Measuring, Analyzing, and Controlling Instruments: Photographic Medical and Optical Goods; Watches and Clocks	0.1928 (0.0635)	3.04	0.1 (0.0514)	1.94	0.091713 (0.0504)	1.82
Business Services	-0.0717 (0.0931)	-0.77	0.00169 (0.0582)	-0.03	-0.0797 (0.0746)	-1.07

*Standard errors in parenthesis

Another industry whose R&D expenditures had a large effect on firms' values is the Mechanical Goods industry. In 2006, for every one percent increase in the percentage change in R&D expenditures to net sales, the price to book value increased by nearly 0.2 percent for a given firm in this industry. The Electronics industry price to book value increased by 0.15 percent and 0.12 percent in 2005 and 2004, respectively.

The Chemicals industry and the Business Services industry did not produce statistically significant results. These two industries produced low coefficient estimates in nearly every year of the study. One can say that because of the low coefficient estimates and the insignificant t-statistics, the Chemicals industry and the Business Services industry does not reward a firm for R&D expenditures in a significant manner.

These results suggest that the relationship between R&D expenditures and market value are industry specific. While we limited our analysis to five industry classifications, we should note that the industries that are typically associated with rapid technological improvements benefit from greater R&D expenditures. Industries with more mature activity or that are service-oriented may not reap as much benefit from greater R&D expenditures.

Other studies that focus on R&D expenditures have shown a positive relationship between government and private sector R&D expenditures and economic growth (BJK Associates 2002).

Klette and Grilliches (2000) found a positive relationship between firm growth and R&D expenditures. While these results show the effects that R&D expenditures have on the economy and firm growth, our results are consistent with the theoretical suggestions that R&D expenditures have a positive impact on firms. The impact that we examine is on firm value.

6. CONCLUSION

We investigated certain drivers and outcomes of innovative activity among small businesses, which comprise a large part of the U.S. economy. Our analysis of a dataset compiled from large databases demonstrated that increasing employment within small businesses enhances innovative activity, while increases in small business sales do not. There may be many conclusions that we can draw from this interrelationship, including further investigations into how policy should be shaped concerning employment growth versus sales growth when focusing on innovative activity.

We also found that increases in R&D activities enhance the value of small businesses in certain industries. It would seem that R&D expenditures account for a wide assortment of innovative activities and are better measures for capturing value-creating innovative activities than more pinpointed statics such as patent-count.

This paper also leads to many additional topics that would benefit from future research. Such topics include understanding what other industries benefit from enhanced R&D expenditures, finding other metrics to measure innovation, and the relationship between investment in innovative activities and the success of small businesses.

Such questions are beyond the scope of this paper, yet an understanding of such questions would likely be fruitful areas of research to benefit those interested in how small businesses innovate and how innovation affects small businesses.

APPENDIX 1. INDUSTRY DESCRIPTIONS

Chemicals and Allied Products

This major group includes establishments producing basic chemicals and establishments manufacturing products by predominantly chemical processes. Establishments classified in this major group manufacture three general classes of products: (1) basic chemicals, such as acids, alkalies, salts, and organic chemicals; (2) chemical products to be used in further manufacture, such as synthetic fibers, plastics materials, dry colors, and pigments; and (3) finished chemical products to be used for ultimate consumption, such as drugs, cosmetics and soaps; or to be used as materials or supplies in other industries, such as paints, fertilizers, and explosives. The mining of natural alkalies and other natural potassium, sodium, and boron compounds, of natural rock salt, and of other natural chemicals and fertilizers are classified in Mining, Industry Group 147. Establishments primarily engaged in manufacturing nonferrous metals and high-percentage ferroalloys are classified in Major Group 33; those manufacturing silicon carbide are classified in Major Group 32; those manufacturing baking powder, other leavening compounds, and starches are classified in Major Group 20; and those manufacturing artists' colors are classified in Major Group 39. Establishments primarily engaged in packaging, repackaging, and bottling of purchased chemical products, but not engaged in manufacturing chemicals and allied products, are classified in Wholesale or Retail Trade industries.

INDUSTRIAL AND COMMERCIAL MACHINERY AND COMPUTER EQUIPMENT

This major group includes establishments engaged in manufacturing industrial and commercial machinery and equipment and computers. Included are the manufacture of engines and turbines; farm and garden machinery; construction, mining, and oil field machinery; elevators and conveying equipment; hoists, cranes, monorails, and industrial trucks and tractors; metalworking machinery; special industry machinery; general industrial machinery; computer and peripheral equipment and office machinery; and refrigeration and service industry machinery. Machines powered by built-in or detachable motors ordinarily are included in this major group, with the exception of electrical household appliances. Power-driven handtools are included in this major group, whether electric or otherwise driven.

ELECTRONIC AND OTHER ELECTRICAL EQUIPMENT AND COMPONENTS, EXCEPT COMPUTER EQUIPMENT

This major group includes establishments engaged in manufacturing machinery, apparatus, and supplies for the generation, storage, transmission, transformation, and utilization of electrical energy. Included are the manufacturing of electricity distribution equipment, electrical industrial apparatus, household appliances, electrical lighting and wiring equipment, radio and television receiving equipment, communications equipment, electronic components and accessories, and other electrical equipment and supplies.

Measuring, Analyzing, and Controlling Instruments; Photographic, Medical, and Optical Goods; Watches and Clocks

This major group includes establishments engaged in manufacturing instruments (including professional and scientific) for measuring, testing, analyzing, and controlling, and their associated sensors and accessories; optical instruments and lenses; surveying and drafting instruments; hydrological, hydrographic, meteorological, and geophysical equipment; search, detection, navigation, and guidance systems and equipment; surgical, medical, and dental instruments, equipment, and supplies; ophthalmic goods; photographic equipment and supplies; and watches and clocks.

Business Services

This major group includes establishments primarily engaged in rendering services, not elsewhere classified, to business establishments on a contract or fee basis, such as advertising, credit reporting, collection of claims, mailing, reproduction, stenographic, news syndicates, computer programming, photocopying, duplicating, data processing, services to buildings, and help supply services.

APPENDIX 2. STANDARD & POOR'S COMPUSTAT – DEFINITIONS OF RELEVANT INFORMATION

1.1. Employees

Mnemonic: EMP
Units: Thousands

This item represents the number of company workers as reported to shareholders. This is reported by some firms as an average number of employees and by some as the number of employees at year-end. No attempt has been made to differentiate between these bases of reporting. If both are given, the year-end figure is used. This item, for banks, always represents the number of year-end employees.

3. Shipping companies' operating differential subsidies and income on reserve fund securities when shown separately;
4. Finance companies' earned insurance premiums and interest income for finance companies, the sales are counted only after net losses on factored receivables purchased;
5. Airline companies, net mutual aid assistance and federal subsidies;
6. Cigar, cigarette, oil, rubber, and liquor companies' net sales are after deducting excise taxes;
7. Income derived from equipment rental is considered part of operating revenue;
8. Utilities' net sales are total current operating revenue;
9. For banks, this item includes total current operating revenue and net pretax profit or loss on securities sold or redeemed;
10. Life insurance, and property and casualty companies' net sales are total income;
11. Advertising companies' net sales are commissions earned, not gross billings;
12. Franchise operations' franchise and license fees;
13. Leasing companies' rental or leased income;
14. Hospitals' sales net of provision for contractual allowances (will sometimes include doubtful accounts); and
15. Security brokers' other income.

This item excludes:
1. Nonoperating income;
2. Interest income (included in Nonoperating Income [Expense]);
3. Equity in earnings of unconsolidated subsidiaries (included in Nonoperating Income [Expense]);
4. Other income (included in Nonoperating Income [Expense]);
5. Rental income (included in Nonoperating Income [Expense]);
6. Gain on sale of securities or fixed assets (included in Special Items);
7. Discontinued operations (included in Special Items);
8. Excise taxes (excluded from sales and also deducted from Cost of Goods Sold); and
9. Royalty income (included in Nonoperating Income [Expense]).

This item includes:

1. All part-time and seasonal employees; and
2. All employees of consolidated subsidiaries, both dom foreign.

This item excludes:

1. Contract workers;
2. Consultants; and
3. Employees of unconsolidated subsidiaries.

1.2. Sales-Net

Mnemonic: SALE
Units: Millions of dollars

This item represents gross sales (the amount of actual bill customers for regular sales completed during the period) reduced discounts, trade discounts, and returned sales and allowances for which is given to customers.

This item is scaled in millions. For example the 1999 annual sales are 173215.000 (or 173 billion, 215 million dollars).

This item includes:

1. Any revenue source that is expected to continue for the life company;
2. Other operating revenue;
3. Installment sales; and
4. Franchise sales (when corresponding expenses are available).

Special cases (by industry) include:

1. Royalty income when considered operating income (such as companies, extractive industries, publishing companies, etc.);
2. Retail companies' sales of leased departments when correspon costs are available and included in expenses (if costs are not avail the net figure is included in Nonoperating Income [Expense]);

1.3. Research and Development Expense

Mnemonic: XRD
Units: Millions of dollars
This item represents all costs incurred during the year that relate to the development of new products or services. This amount is only the company's contribution.

This item includes:

1. Software expenses; and
2. Amortization of software costs

This item excludes:

1. Customer or government-sponsored R&D (including reimbursable indirect costs);
2. Extractive industry activities, such as prospecting, acquisition of mineral rights, drilling, mining, etc.;
3. Engineering expense—routine, ongoing efforts to define, enrich, or improve the qualities of existing products;
4. Inventory royalties; and
5. Market research and testing.

This item is not available for banks and utilities.

1.4. Price to Book

Mnemonic: MKBK
Units: Percentage
Price to Book Ratio is defined as Market Value - Monthly divided by Quarterly Common Equity - Total, which represents the common shareholder's interest in the company, including common stock, capital surplus, retained earnings and treasury stock adjustments. (If Common Equity for the current quarter is not available, the values for the previous quarter will be used.)

1.5. Price to Book Fiscal Year End

Mnemonic: MKBKF
Units: Percentage
Price to Book Ratio - Fiscal Year-End is Market Value - Fiscal Year-End (or the close price for the fiscal year) multiplied by the company's common shares outstanding, divided by Common Equity as Reported, which represents the common shareholders' interest in the company.

APPENDIX 3. REFERENCES

Amemiya, T. (1985). *Advanced Econometrics.* Cambridge: Harvard University Press.

Arrellano, M. & Bond, S. (1991). "Some tests of specification for panel data: Monte Carlo evidence and an application to employment equations," *Review of Economic Studies, 58,* 277-297.

Audretsch, D. & Acs, Z. (1998). "Innovation in Small and Large Firms: An Empirical Analysis" *American Economic Review, 78*(4), 678-90.

Audretsch, D. & Acs, Z. (1991). "Innovation and Size at the Firm Level," *Southern Economic Journal,* Vol. 57, No. 3, 739-744.

BJK Associates (2002). "The Influence of R&D Expenditures on New Firm Formation and Economic Growth," *Small Business Administration, Office of Advocacy,* Contract Number SBAHQ-00-M-0491.

Blundell, R., Griffith, R. & Van Reenen, J. (1995). "Dynamic count data models of technological innovation," *Economic Journal, 105,* 333-344.

Bollerslev, T., & Ghysels, E. (1996). "Periodic Auto-regressive Conditional Heteroscedasticity," *Journal of Business and Economic Statistics,* Vol. *14,* 139-151.

Bound, J., Cummins, C., Griliches, Z., Hall, B. H. & Jaffee, A. (1984). "Who Does R&D and Who Patents in *R&D. " Parents, and Productivity,* edited by Zvi Griliches. Chicago: University of Chicago, 21-54.

Breitzman, Anthony & Hicks Diana (2008). "An Analysis of Small Business Patents by Industry and Firm Size," *Small Business Administration, Office of Advocacy,* Contract Number SBAHQ-07- Q-0010.

Cincer, M. (1997). "Patents, R&D, and Technological Spillovers at the Firm Level: Some Evidence from Econometric Count Models for Panel Data," *Journal of Applied Econometrics, Vol. 12,* No. 3, 265-280.

Coulson, N. & Robins, R. (1985) "Aggregate Economic Activity and the Variance of Inflation: Another Look," *Economics Letters, Vol. 17*, 71-75

Cragg, J. (1982). "Estimation and Testing in Testing in Time Series Regression Models with Heteroscedastic Disturbances," *Journal of Econometrics, Vol. 20*, 135-157.

Crepon, B. & E. Duguet (1997). "Estimating the Innovation Function from Patent Numbers: GMM on Count Panel Data," *Journal of Applied Econometrics*, this issue.

Davidson, R. & Mackinnon J. (1993). *Estimation and Inference in Econometrics.* New York: Oxford University Press.

Domowitz, I. & Hakkio, C. (1985). "Conditional Variance and the Risk Permium in the Foreign Exchange Market," *Journal of International Economics, Vol. 19*, 47-66.

Engle, R. (1982) "Autoregressive Conditional Heteroscedasticity with Estimates of the Variance of United Kingdom Inflations," *Econometrica, Vol. 50*, 987-1008.

Engle, R. Hendry, D & Trumble, D. "Small Sample Properties of ARCH Estimators and Tests," *Canadian Journal of Economics, Vol. 18*, 66-93.

Gourieroux, C., Monfort, A. & Trognon, A. (1984a). "Pseudo Maximum Likelihood Methods: Theory," *Econometrica, 52*, 68 1-700.

Greene, William A. (2003). *Econometric Analysis: Fifth Edition.* Pearson Education, Inc. Upper Saddle River.

Griliches, Z. (1979). "Issues in Assessing the Competition of R&D to Productivity Growth," *Bell Journal of Economics, 10*, 92-1 16.

Hausman, J., Hall, B. H. & Griliches, Z. (1984). "Econometric models for count data with an application to the patents-R&D relationship," *Econometrica, 52*, 909-938.

Hausman, J., Hall, B. H. & Griliches, Z. (1986). "Patents and R&D: Is There a Lag?" *International Economic Review, 27*, 265-283.

Jaffe, A. B. (1986). "Technological opportunity and spillovers of R&D," *American Economic Review, 76*, 984-1001.

Jaffe, A. B. (1988). "R&D intensity and productivity growth," *Review of Economics and Statistics, 70*,43 1-437.

Judge, G, Griffiths, W., Hill, C. & Lee, T. (1985). *"The Theory and Practice of Econometrics,"* New York: John Wiley and Sons

Lerner J. (1994). "The Importance of Patent Scope: *An Empirical Analysis,"* Vol. 25, No. 2, 319- 333

Loury, G. (1979). "Market Structure and Innovation," *Quarterly Journal of Economics, 93*, 395- 410.

Lucas, R. (1978). "On the Size Distribution of Business Firms," *Bell Journal of Economics, 9(2)*, 508.

Mairesse, J. & Hall, B. H. (1996). "Estimating the Productivity of Research and Development: An Exploration of GMM Methods Using Data on French and United States Manufacturing Firms," *NBER Working Paper 5501*, National Bureau of Economic Research, Cambridge.

Pakes, A, & Griliches, Z. (1984). "Patents and R&D at the Firm Level: A First Look," Z. Griliches (ed.). *R&D, Patents and Productivity*, University of Chicago Press, Chicago.

Perline, R, Axtell, P, & Teitelbaum, D. (2006). "Volatility and Asymmetry of Small Firm Growth Rates Over Increasing Time Frames," *Small Business Administration, Office of Advocacy*, No. 285, Contract No. SBAHQ-05-Q-00 18,

Pianta, M. & Vaona, A. (2006). "Innovation and Productivity in European Industries," *Kiel Working Papers 1283*, Kiel Institute for the World Economy.

Rogers, M. (2000). "Understanding Innovative Firms: An Empirical Analysis of the GAPS," *Melbourne Institute Working Paper Series* wp2000n08, Melbourne Institute of Applied Economic and Social Research, The University of Melbourne.

Scherer, F. M. (1965). "Firm Size, Market Structure, Opportunity, and the Output of Patented Inventions," *American Economic Review*, LV , 1097-1125. (b)

Scherer, F. M. (1983). "The Propensity to Patent," *International Journal of Industrial Organization*, 107-28.

Scherer, F. M. (1984). *Innovation and Growth: Schumpeterian Perspectives.* Cambridge, Mass.: MIT Press.

Scherer, F. M. (1991). "Changing Perspectives on the Firm Size Problem", *Innovation and Technological Change: An International Comparison*, edited by Zoltan J. Acs and David B. Audretsch. Ann Arbor: University of Michigan Press, 1991.

Schumpeter, J. A. (1934). *Theory of economic development.* Cambridge: Harvard University Press.

Van Dijk, B. et al, 1997. "Some New Evidence on the Determinants of Large- and Small-Firm Innovation," *Small Business Economics*, Springer, vol. 9(4), pages 335-43, August.

WallSten, S. (2000). "The Effects of Government-Industry R&D Programs on Private R&D: The Case of the Small Business Innovation Research Program," *The RAND Journal of Economics, Vol 31*, No 1, 82-100.

Winkelmann, R, & Zimmerman, K. F. (1995). "Recent Developments in Count Data Modeling: Theory and Application," *Journal of Economic Surveys, 9,* 1-24.

End Notes

[1] Appendix 1 has a full description of each industry.
[2] U.S Small Business Administration website:
http://app1.sba.gov/faqs/faqIndexAll.cfm?areaid=24, as of June 22, 2007.
[3] U.S Small Business Administration website:
http://app1.sba.gov/faqs/faqIndexAll.cfm?areaid=24, as of June 22, 2007.
[4] Mnemonics are short codes that allow a user to easily pull and analyze data contained within Compustat database.
[5] We contacted the USPTO to confirm that this was the most appropriate method by which to gather data from the database.
[6] There may be a "chicken and the egg" problem that is not explored in this paper: Do revenues create opportunities for more innovation, or does innovation create more opportunities for revenue? We have focused on the former causality at this juncture.
[7] Engle (1982, 1983) and Cragg (1982) found evidence that for some kinds of data, the disturbance variances in time-series models were less stable than usually assumed. Engle's results "suggested that in models of inflation, large and small forecast errors appeared to occur in clusters, suggesting a form of heteroscedasticity in which the variance of the forecast error depends on the size of the previous disturbance" (Greene 2000).
[8] Other research that used the moving average model include Coulson and Robins (1985), Engle, Hendry, and Trumbull (1985), Domowitz and Hakkio (1985), and Bollerslev and Ghysels (1996).
[9] Literature shows that public and private information is fully embedded in market value, i.e. stock prices. Efficiency within a firm is included when valuing a firm. References for this topic include Harris & Gurel (1986) and Mitchell & Mulherin (1994).
[10] After performing analysis to measure the effects of the lag variables, the results produced were inefficient and did not provide information on which lag structure would be optimal.
[11] A more detailed discussion of how the functional form corrects for heteroscedasticity and autocorrelation is found in Judge et al. (1985), Amemiya (1985), and Davidson and MacKinnon (1993).
[12] The lags of patent counts were not used because research by Mitchell and Mulherin et al. (1994) shows that markets are very efficient. Public and private information instantly impacts the value of a firm. Because of these findings, we believe that the current year's patent production would most immediately impact the price to book value for the given year. Previous years' patent count would be reflected in its respective years.
[13] For this model, we explored other possible determinants of the price-to-book ratio. These determinants included net sales revenues, operating profit, research and development expenses, earnings before depreciation, shares outstanding, and closing stock price for the year. We found that there were no significant relationship between any of the above determinants and the price-to-book ratio.
[14] Similar to equation (6), we applied the same determinants to this equation. In our results, none of the determinants had any significant effect on the model.
[15] While we are measuring for employee headcount and revenues, it is necessary to include the other variables because they provide a better fit for the model, giving consistent and efficient estimates.

In: Innovators and Tinkerers Role in Business: A... ISBN: 978-1-60876-815-8
Editors: Steven C. Hymes © 2010 Nova Science Publishers, Inc.

Chapter 2

NETWORK OF TINKERERS: A MODEL OF OPEN-SOURCE TECHNOLOGY INNOVATION

Peter B. Meyer[*]

ABSTRACT

Airplanes were invented by hobbyists and experimenters, and some personal computers were as well. Similarly, many open-source software developers are interested in the software they make, and not focused on profit. Based on these cases, this paper has a model of agents called tinkerers who want to improve a technology for their own reasons, by their own criteria, and who see no way to profit from it. Under these conditions, they would rather share their technology than work alone. The members of the agreement form an information network. The network's members optimally specialize based on their opportunities in particular aspects of the technology or in expanding or managing the network. Endogenously there are incentives to standardize on designs and descriptions of the technology. A tinkerer in the network who sees an

---

[*] With thanks to Harley Frazis, Tomonori Ishikawa, Anastasiya Osborne, Larry Rosenblum, Leo Sveikauskas, Cindy Zoghi, and participants at seminars at BLS, the Midwest Economics Association, BEA, the Naval Postgraduate School, the 2006 International Economic History Congress, and OSSEMP 2007. JEL code O31. Keywords: Open source; invention; innovation; tinkerers; technological change

opportunity to produce a profitable product may exit the network to create a startup firm and conduct focused research and development. Thus a new industry can arise.

Some important technologies have been advanced by open sharing among innovators who were not motivated mainly by prospective profits. For example, many hobbyists around the world tried to make aircraft in the late 1800s, before there were what we now call airplanes. Personal computers were advanced greatly by hobbyists who met in groups, notably at the Homebrew Computer Club (Freiberger and Swaine (1984) and Levy (2001)). Many open-source software projects make source code publicly available. The airplane, personal computer, and open-source software cases are examples of "open source" technology development processes.

Allen (1983) introduced the related term *collective invention* to describe firms sharing technical information. Schrader (1991), Nuvolari (2002), and von Hippel (2005) offer other examples. Harhoff, Henkel, and von Hippel (2003) model this phenomenon. In this literature, the technology is known to deliver useful outputs, and the profit-minded firms exchange information. By contrast, this paper describes situations in which a novel technology appears *first* because of the combined efforts of people who do not expect to sell anything.

This paper places open source technology development in an abstract, deductive model. Three key assumptions are necessary. First, agents called tinkerers are interested in advancing the technology for some reason, such as their own inherent interest. Second, each tinkerer sees how to improve the technology using his own criteria for improvement. Third, the tinkerers believe the technology is so immature and uncertain that current actions do not significantly affect future opportunities for commercialization. Under these conditions, tinkerers share their technologies with one another, forming an an open-source network.

Within the model, a new industry appears when a tinkerer envisions a way to profit from the technology, and leaves the network to try that. In the cases described, amateur tinkerering eventually led to increases in commercial output and productivity.

I. EXAMPLES OF OPEN-SOURCE TECHNOLOGY DEVELOPMENT

I.A. Before the Airplane

For decades before there were functional airplanes, there was an international discussion about wings and flying machines. By the 1890s several journals and societies in France, Britain, Germany, and the United States were devoted to this topic. Important experiments by Otto Lilienthal, Samuel Langley, and Lawrence Hargrave advanced the field.

A Chicago railroad engineer named Octave Chanute was inspired by the possibility that by cooperating, experimenters around the world could make winged flying machines a reality. He visited many of them, and corresponded with many more. Chanute's speeches and writings were "noteworthy for fostering a spirit of cooperation and encouraging a free exchange of ideas among the world's leading aeronautical experimenters" (Stoff, 1997, p. iv). In his optimistically titled 1894 book *Progress in Flying Machines*, Chanute summarized and commented on hundreds of kites, gliders, experimenters, authors, and theorists of aerial navigation. Newly interested people learned about the subject from this important book. Wilbur and Orville Wright read it and contacted Chanute.

Like many others, the Wrights discussed their experiments openly. Chanute visited them and invited colleagues to participate in their effort. At Chanute's invitation, Wilbur Wright made a speech about their experiments at the Western Society of Engineers. Wilbur Wright published in British and German aircraft journals. In other words, the Wrights took an open source perspective on their technology as they advanced it.

In 1902 and 1903, the Wrights developed better wings and propellers than their predecessors, partly because of the uniquely accurate and precise measurements they got from their wind tunnel and its instrumentation. They began to withdraw from processes of open sharing as they believed they were near to a successful powered glider flight. (Crouch, 2002). They planned to protect their rights to patent and license their technology. This led to permanent conflicts with Chanute, who was devoted to open-source processes of invention.

I.B. The Beginning of Personal Computers

In the 1970s many clubs of hobbyists were working on microcomputers. The Homebrew Computer Club which met in Menlo Park and Palo Alto, California, starting in March, 1975 was particularly central. Most of the people who attended were interested in making computers for their own home use. At the first meeting, "it turned out that six of the thirty-two had built their own computer system of some sort, while several others had ordered Altairs" (Levy, 2001, p. 202). The Altair was a new kit for making a hobbyist computer.

Meetings were informal. "The group had no official membership, no dues, and was open to everyone. The newsletter, offered free ... became a pointer to information sources and a link between hobbyists." (Freiberger and Swaine, 1984, p. 106) "They discussed what they wanted in a club, and the words people used most were 'cooperation' and 'sharing'." (Levy, p. 202). Homebrew meetings included a presentation, often of a demonstration of a club member's latest creation. Then there was "the Random Access session, in which everyone scrambled around the auditorium to meet those they felt had interest in common with them....[M]uch information had to be exchanged; they were all in unfamiliar territory" (Freiberger and Swaine, 1984, p. 106).

The information flow was a cause and also an effect of the fact that they often used similar parts, attempted similar projects, and read the same newsletters and magazines. Members were drawn to the hands-on experience of making computers and understanding the component parts, not theories of computing, or the social effects of computing. (Levy, 2001)

The Homebrew club of hobbyists had an important effect in moving personal computer technology forward. There were many other places for hobbyists to get involved in this exciting area. There were a series of (U.S.) West Coast Computer Faires which gathered tremendous interest and attendance. Hobbyists ran bulletin board personal computer systems to which people could dial in and send email, and engaged in Usenet discussions on the Internet. Hobbyists did this activity, mostly not for profit.

At one Homebrew meeting, Steve Wozniak demonstrated a new board which could do many things a computer would do. He did not intend to start a company or sell anything, but his entrepreneurial friend Steve Jobs convinced him to cofound a company and to sell this product as a computer, which they called the Apple I. Only computer hobbyists could use it, but among them it was quickly in demand. The personal computer industry took off with this device.

Apple Computer, and perhaps twenty other companies, were started by Homebrew attendees. But the club started because of an interest in computers, not business.

I.C. Open Source Software Projects

In open-source software, human-readable source code files are made widely available on a computer network. Source code, in computer languages, is fed as input to specialized development tool programs, such as interpreters, compilers, assemblers, and linkers, which generate the instructions which a computer eventually executes.

Sharing source code makes it possible for many programmers to experiment and improve the code in parallel. A user may also alter the program for a particular purpose. Sponsors of open source projects usually copyright the software in a way that allows a wide spectrum of uses. Revisions are published under the same license. This is a powerful mechanism to support collective invention because it is common knowledge that some later improvements will become part of the shared code.

The owners of a chunk of source code moderate the final choices in released versions of the software. Users may make a version different from a released one. The owners try to avoid the project's source code "forking" into permanently divergent, partly-incompatible versions. If that were to happen, the project's members would lose some of the benefits of having one code base which improved along many dimensions over time.

Several roles and institutions support sharing in open source projects:

- Web servers store the source code.
- Intellectual property claims are explicitly preempted by special open copyrights.
- The relevant programmers have similar development tools and skills.
- Source control programs keep records of who changed the software and how.
- Moderators or "owners" control which of those changes are published.
- Culturally, experimentation is welcome and unrestricted.

Such projects have been started by individuals with many different interests. The operating system Linux, for example, was started, sponsored, and organized by a student, Linus Torvalds. Now it is a core product of firms with hundreds of millions of dollars in revenue annually. Many other projects such as Apache and Firefox also have this form. Open source software projects often have an explicit copyright condition to keep the core technology in the public domain.

II. MOTIVATION AND PSYCHOLOGY

The model which follows is meant to describe the airplane experimenters and also hobbyists, hackers, and innovators of the computer age such as Steve Wozniak (developer of the Apple I personal computer), Richard Stallman (a defining programmer of the free-software movement), Tim Berners-Lee (inventor of the World Wide Web's browsers and servers), and Linus Torvalds (the founding programmer of the Linux operating system). These innovators created important technologies without intending to sell them.

Such innovators have various motivations. They may find a project inherently absorbing and enjoyable. They may benefit from some service it provides. These are sometimes called *intrinsic* motivations. They may anticipate receiving honors, prestige, wealth, or career benefits from the project, which are *extrinsic* motivations. They may anticipate that the project could improve the human condition apart from themselves, which is an *altruistic* motivation. The model to follow directly incorporates intrinsic or altruistic motivations, and demonstrates how certain network behaviors emerge.

Important aircraft experimenters referred to their intrinsic or altruistic motivations:

- "A desire takes possession of man. He longs to soar upward and to glide, free as the bird . . ." (Otto Lilienthal, 1889).
- "The glory of a great discovery or an invention which is destined to benefit humanity [seemed] . . . dazzling. . . . Otto and I were amongst those [whom] enthusiasm seized at an early age." (Gustav Lilienthal, 1912 introduction).
- "The writer's object in preparing these articles was ... [to ascertain] whether men might reasonably hope eventually to fly through the air

- ... [and] To save ... effort on the part of experimenters ..." (Chanute, 1894).
- "I am an enthusiast ... as to the construction of a flying machine. I wish to avail myself of all that is already known and then if possible add my mite to help on the future worker who will attain final success" (from Wilbur Wright's 1899 letter to the Smithsonian Insitution requesting information).
- "Our experiments have been conducted entirely at our own expense. At the beginning we had no thought of recovering what we were expending, which was not great ..." (Orville Wright, 1953, p. 87).

The motivation of hardware hackers was often instrinsic or altruistic too. After the first meeting of the Homebrew Club, Steve Wozniak reports (Wozniak, 2006, pp 156-7):

> I started to sketch out on paper what would later come to be known as the Apple I. . . . I did this project for a lot of reasons. For one thing, it was a project to show the people at Homebrew that it was possible to build a very affordable computer . . . with just a few chips. In that sense, it was a great way to show off my real talent, my talent of coming up with clever designs, designs that were efficient and affordable. By that I mean designs that would use the fewest components possible.
>
> I also designed the Apple I because I wanted to give it away for free to other people. I gave out schematics for building my computer at the next meeting I attended.
>
> This was my way of socializing and getting recognized. I had to build something to show other people. And I wanted the engineers at Homebrew to build computers for themselves...

Open source developers have a similar mix of motives. Lakhani and Wolf (2003) show based on surveys that many programmers participate in open source projects because of the creative enjoyment and the value of using the output, not explicit rewards. **Pavlicek (p. 146) reports that "Open Source people are used to doing work on a project because they perceive its value to the community."**

It is difficult to define in output or engineering terms what the tinkerers, hobbyists, or hackers are accomplishing in the short run. The devices or software do not work well, and they are not clearly commensurable, because they are qualitatively different attempts to make a desirable design.

In the model to follow, progress is therefore not measured by attributes of the artifacts, but by the individual's own satisfaction with it, that is, in terms of utility.

III. A TINKERER

Let us define an individual called a tinkerer who enjoys a technological activity A. The notation A stands for aircraft or some other hobbyist activity such as building a computer or writing a computer program at home. *A* has no market value, and no honors or profits are associated with it. The tinkerer may imagine that there may someday be honors or profits, but thinks this is unlikely and assigns a low expected value to such possibilities.

The tinkerer receives a periodic flow of positive utility at directly from the existence of A in period *t*. Let $a_0 \geq 1$ be a parameter defining the utility received in period zero, the present period, and treat the choice about tinkering separately from all other utility decisions. The tinkerer values alternative choices in a risk-neutral way according to the net present sum of expected future utility payoffs. Utility to be received in future periods is discounted by a $\beta \in (0, 1)$ for each intervening period. Using a standard time series summation $((1-\beta)(\sum_{t=0}^{\infty} \beta^t) = 1)$, expected utility at time zero can be put into a closed form:

$$EU_{t=0} = a_0 + \beta a_0 + \beta^2 a_0 + \cdots = a_0 \sum_{t=0}^{\infty} \beta^t = a_0 (\sum_{t=0}^{\infty} \beta^t)(\frac{1-\beta}{1-\beta}) = \frac{a_0}{1-\beta}$$

(1)

The tinkerer can choose to invest in ("tinker with") *A* in order to raise future benefits a_t. An investment costs one utility unit in the present period representing the effort, expenses, and the opportunity costs of time involved. The agent anticipates that tinkering will raise his future utility by *p* units in each time period in the future. The notation *p* stands for progress which the agent experiences subjectively. We assume *p* is fixed, positive, and that the tinkerer's forecast is correct.

A tinkerer chooses whether to tinker based on the estimated costs and benefits. The utility benefits from one effort to tinker have the value *p* in each subsequent period. The discounted payoffs to tinkering in the present period are

$$p\beta + p\beta^2 + p\beta^3 + p\beta^4 + \ldots = \frac{p\beta}{1-\beta}$$

The investment required to receive this payoff is one utility unit at time zero so the net payoff to tinkering in period zero is $\frac{p\beta}{1-\beta} - 1$. Benefits exceed cost when $p > \frac{1-\beta}{\beta}$. For example, if $\beta = 0.95$ and $p = 0.07$, tinkering in the current period brings a positive surplus of $\frac{.07*.95}{.05} - 1 = .33$.

Unless parameters or conditions change, any tinkerer who finds it worthwhile to tinker once will find it worthwhile to tinker again and again. As long as $p > \frac{1-\beta}{\beta}$, the agent will tinker in every period, receiving payoff of $a_0 - 1$ in the current period, $a_0 + p - 1$ in period one, and in general $a_0 + pt - 1$ in period t. The associated payoff stream is

$$EU_{t=0} = \sum_{t=0}^{\infty} \beta^t (a_0 + pt - 1)$$

$$= (a_0 - 1) \sum_{t=0}^{\infty} \beta^t + p \sum_{t=0}^{\infty} \beta^t t$$

$$= \frac{a_0}{1-\beta} - \frac{1}{1-\beta} + p \sum_{t=0}^{\infty} \beta^t t$$

The last term can be expressed in closed form using this derivation:

$$\sum_{t=0}^{\infty} \beta^t t = \beta + 2\beta^2 + 3\beta^3 + \cdots$$

$$= (\beta + \beta^2 + \beta^3 + \cdots) + (\beta^2 + \beta^3 + \beta^4 + \cdots) + (\beta^3 + \beta^4 + \beta^5 + \cdots) + \cdots$$

$$= \frac{\beta}{1-\beta} + \beta \frac{\beta}{1-\beta} + \beta^2 \frac{\beta}{1-\beta} + \beta^3 \frac{\beta}{1-\beta} + \cdots$$

$$= \frac{\beta}{1-\beta} (1 + \beta + \beta^2 + \beta^3 + \cdots)$$

$$= \frac{\beta}{1-\beta} \left(\frac{1}{1-\beta} \right)$$

$$= \frac{\beta}{(1-\beta)^2}$$

With that substituted in, the tinkerer's overall expected utility at time zero is:

$$EU_{t=0} = \frac{a_0}{1-\beta} - \frac{1}{1-\beta} + \frac{p\beta}{(1-\beta)^2} \qquad (2)$$

The first term of equation 2 expresses the present value of expected utility from A in its original state. The second term is the present value of the costs of endless tinkering. The third term is the present value of the benefits expected from endless tinkering.

For a tinkerer characterized by $\beta = 0.95$ and $p = 0.07$, the gain in expected utility from tinkering forever is the sum of the second and third terms:

$$\frac{p\beta}{(1-\beta)^2} - \frac{1}{1-\beta} = \frac{.07 * .95}{(0.05)^2} - 20 = \frac{1.33}{.05} - 20 = 6.6$$

So, for these parameters (which will be used throughout the paper to facilitate comparison), endless tinkering increases the tinkerer's utility by 6.6 times the cost of a one-time investment. This self-motivated tinkerer is a perpetual innovation machine.

IV. A NETWORK OF TINKERERS

To get to the main proposition quickly, we make simple and extreme assumptions. Later sections relax the underlying assumptions.

Let there be two tinkerers with identical utility functions working on similar projects A_1 and A_2 whose innovative tinkerings could be useful to one another. Each tinkerer believes that the other cannot profit from the project using any foreseeable version of the existing technology. Let the subjective rate of progress of the first player be p_1, and the subjective rate of progress of player two be p_2.

Suppose the two tinkerers can make a verifiable and enforceable agreement to share a well-defined set of the functional design changes in A_1 and A_2 and their experimentally discovered effects. This agreement forms a *network* for future information. At any time, either partner can depart from the network, and then ceases to share his subsequent innovations and ceases to learn from the other tinkerer.

Let fraction $f \in (0, 1)$ of each tinkerer's innovation be perceived as useful to the other one's project, so that knowing tinkerer two's most recent innovation would benefit tinkerer one by fp_2 each turn. The remaining fraction $(1 - f)$ does not carry over because the projects are not identical and perhaps there are costs to interacting.

If the tinkerers expect each other to produce a positive flow of innovations, they are always better off by joining in a network. If they tinker and share with these parameters forever, tinkerer one's expected utility is:

$$EU_0 = \frac{a_0}{1-\beta} - \frac{1}{1-\beta} + \frac{p_1\beta}{(1-\beta)^2} + \frac{fp_2\beta}{(1-\beta)^2} \tag{3}$$

The new fourth term expresses the benefits player one receives from the flow of information coming from player two. Because of this free good, utility is greater in equation (3) than in equation (2). The tinkerer prefers joining a network rather than working alone. Thus *under these assumptions, rational agents generate open-source technology networks.* This is the central analytical result of this paper.

V. STANDARDIZING, SPECIALIZING, AND CONSENSUS REDESIGN

Only a fraction $f \in (0, 1)$ of the experimental discoveries made by player two are usable to player one. Suppose that for cost c_s, a tinkerer can adjust some elements of his project to look more like the other one's project, and that doing so would raise the fraction of innovations of the other tinkerer which apply to his own project from f to f_2. If tinkerer one pays this cost to standardize on an element of tinkerer two's design, expected utility changes to

$$EU_0 = \frac{a_0}{1-\beta} - \frac{1}{1-\beta} + \frac{p_1\beta}{(1-\beta)^2}4 + \frac{fp_2\beta}{(1-\beta)^2} - c_s + \frac{(f_2-f)p_2\beta}{(1-\beta)^2}$$

Comparing this to equation **3**, a tinkerer would pay the standardization cost if:

$$\frac{\beta p_2 (f_2 - f)}{(1-\beta)^2} > c_s$$

In words, a player benefits more from standardization if, holding other things constant: (a) other tinkerers produce a large flow of innovations p_2; (b) the cost of standardization, c_s, is small; (c) the increase in usable innovations $(f_2 - f)$ is large; and (d) the tinkerer is patient for results (β is close to 1). Intuitively, these are the conditions under which it makes sense for a software developer to replace a working piece of code by a standard library of code written by others.

The same argument can explain why experimenters tend to develop a common technical language to describe their technologies. This can reduce communication costs and also clarify thinking. For example, Wilbur Wright published a journal article (Wright, 1902) asking other experimenters to cease using "angle of incidence" to mean the angle between a wing (or other airfoil) and the ground. The better definition, he argued, was the angle between the airfoil and the flow of air coming at it; the angle with respect to the ground was not relevant. This request was an effort both to improve the thinking processes of other experimenters and to lower frictional losses in communication. In a more important example, Lawrence Hargrave's experiments showed that a box-shaped kite was more stable than a flat kite in a gust of wind. This specialist contribution helped glider flyers standardize on a biplane (two wing) design for gliders.

Both kinds of standardization partly explain why tinkerers would agree to publish their findings. The fewer differences between experiments there are, the lower future communication and adoption costs will be. It benefits player one in communication if his preferred language and concepts are available to both players. If a tinkerer anticipates adopting part of another's tinkerer's technology at some time in the future, he lowers the future cost of that adoption by giving the other tinkerer a chance to use his own technology now. It also means they would be able to compare options for standardization and choose the best one, in the sense of moving the project forward or raising f more. This incentive could be formalized by making $f()$ a decreasing function of a player's own history of making new findings public. An experimenter who publishes more makes it easier for other players to communicate with him

Network of Tinkerers: A Model of Open-Source Technology Innovation 43

or to learn from his design choices. If f is a declining function of the number of findings a player has shared, it partly substitutes for the enforcement of the rule that players should share all their findings. Each one has an incentive to share in order to get the others to learn from his own findings, and to standardize on his own choices (rather than having to pay the costs of standardizing on the choices of others).

A tinkerer may invest in redesign to make the device easier to learn or easier to use, because it represents progress p or makes it easier to exchange information, raising f. This is important in the software context where a project can "fork" (split over time into incompatible versions used by different people) if the contributors do not agree to standardize enough. In the history of Unix there was a painful fork, and programmers refer to this history to convince others to keep projects unified even as they work independently. In this model, they are willing to pay some price to maintain the large network on the project. A redesign to achieve a consensus and avoid forking is therefore rationalized by the flow of future exchanges that are possible if the players can avoid forking.

For $f = 0.5$, $f_2 = 0.54$, $p_1 = 0.07$, and $\beta = 0.95$, the payoff to standardization is $\frac{p_2\beta(f_2-f)}{(1-\beta)^2} = \frac{.07*.95*.04}{.05*.05} = 1.064$. In this illustration, if the cost of the standardization investment were one utility unit, like the cost of a normal investment, it would be just worth undertaking.

Thus standardization and specialization are natural outcomes of exchanging information to develop a technology. They can occur without any necessary reference to competition or market exchange. The network is a search technology which provides the tinkerer with information he values and does not obtain by experiment.

VI. JOINING, SEARCHING, OR MATCHING COSTS

Perhaps there are costs to finding a match partner or joining together once a match is found. Let c_j be the immediate cost in utility terms to a tinkerer for joining a sharing institution or starting one with a known partner. The gross benefits of joining the group are again $\frac{fp_2\beta}{(1-\beta)^2}$, and if c_j is less than this, the tinkerer would prefer to join than to work alone.

So the model predicts the tinkerer joins, ceteris paribus, if (1) costs of joining, c_j, are small enough, (2) the flow of innovations from the others in the group, p_2, is large enough, (3) the innovations are relevant enough to his own project, as measured by f, and (4) the tinkerer's valuation of future events, β, is high enough.

The same comparison applies if c_j is the cost for a tinkerer to search for a network or candidates to join an existing network. This parameter can incorporate the real-world problem that usually few people know a network exists and how to communicate with it. The problem is addressed in the real world by members who write books, edit journals, make speeches, talk about their hobby to outside people, or broadcast emails.

There might be many tinkerers, working in isolation, making almost no progress because they do not share. Here we have a situation in which an information failure alone prevents a Pareto-improving institution from appearing. Probably there are many situations in which tinkerers *would* join a network, but the search costs are such that they do not find one another. If one thinks of tinkerers as a natural resource, institutional attributes of the environment (like the presence of the Internet) affect whether they can find one another and work together and their speed of progress.

An individual tinkerer might specialize in expanding the network, e.g. through speech-making, book-writing or other publicity. Tinkerers who make A easier to learn or easier to use can also lower search costs by making it easier for others to see the virtues of A.

We do see such editor/moderators in the cases of interest:

- Aircraft experimenter and author Octave Chanute had a strong interest in open sharing of information. He expressed affection for the point of view of Lawrence Hargrave, who on principle published all his results and patented nothing, with the idea that this open-source approach would maximize the speed of collective progress.
- In the Homebrew computer club, Lee Felsenstein, who usually moderated the meetings, established a "Random Access" interaction time for people to talk to whoever could help them.
- In the open source software cases, charismatic founders or charismatic projects draw in interest, and the programmers are explicitly and routinely encouraged to share innovations, sometimes by the licensing agreement.

VII. INTELLECTUAL PROPERTY

Some of the innovators discussed preferred to avoid formal intellectual property claims and institutions, such as patents, which might get in the way of using a technology. Pioneering aircraft experimenter Lawrence Hargrave and programmer Richard Stallman are examples. This behavior can be rationalized in this model. Effort devoted to establishing intellectual property rights in an unprofitable technology may not pay off as well as sharing which pushes the technology forward.

One can formally illustrate this. For simplicity, consider a two-tinkerer case. Assume all the utility functions are linear in money and have been normalized to the money metric, and that none of them expect to be make a commercial product. Suppose each tinkerer has property rights to his designs and can charge a price to use the design information he transmits to the network. He may impose a cost c_1 for each information transmission on each network member who makes use of it. With one network partner, a tinkerer receives c_1 times fp_1 in copyright payments, and pays out c_2 times fp_2. This pattern of zero-sum exchanges is profitable to the tinkerers who produce the greatest flow of innovations, but some of the others may find it too expensive enough to stay in the network, which slows overall progress.

More realistically, if there are many partners and frictional costs to defining, managing, or enforcing intellectual property rights, private ownership may bring the tinkerers greater social costs than social benefits. For both reasons, tinkerers in the model are better off and more willing to participate in networking if the rules of the game do not include the definition and protection of intellectual property.

That changes when commercialization to a broader market, beyond the tiny population of tinkerers, is likely. So far the model has not considered the mixed incentives faced by a tinkerer who anticipates selling a product some day. That tinkerer faces a perceived opportunity cost if he does not create a barrier to competition. In one useful example, the Wrights changed their behavior once they believed they were about to invent the airplane.

VIII. ENTREPRENEURIAL EXITS FROM THE NETWORK

Starting in late 1902, after they had run tests on wings in a wind tunnel, the Wrights were decreasingly willing to share information. From Crouch (2002), p. 296:

> The brothers had been among the most open members of the community prior to this time. The essentials of their system had been freely shared with Chanute and others. Their camp at Kitty Hawk had been thrown open to those men who they had every reason to believe were their closest rivals in the search for a flying machine. This pattern changed after fall 1902.
> The major factor leading to this change was the realization that they had invented the airplane. Before 1902 the Wrights had viewed themselves as contributors to a body of knowledge upon which eventual success would be based. The breakthroughs accomplished during the winter of 1901 and the demonstration of... success on the dunes in 1902 had changed their attitude.

They applied for a patent in March 1903, received it in 1906, and started an airplane business. Chanute had criticized others who kepts secrets before, and he began to have conflicts with the Wrights. These conflicts grew severe and in the end, Chanute and the brothers were no longer on speaking terms.

This kind of split also occurred in the Homebrew club, whose attendees had tended to follow what Levy (2001) called the Hacker Ethic – that information should be freely available. After Apple and other companies were founded by its members, the experience at the club changed. Members who had started companies stopped coming, partly because keeping company secrets would be uncomfortable at Homebrew. From Levy (2001), p. 269:

> No longer was it a struggle, a learning process, to make computers. So the pioneers of Homebrew, many of whom had switched from building computers to *manufacturing* computers, had not a common bond, but competition to maintain market share. It retarded Homebrew's time-honored practice of sharing all techniques, of refusing to recognize secrets, and of keeping information going in an unencumbered flow. . . . Now, as major shareholders of companies supporting hundreds of employees, they had secrets to keep.
> "It was amazing to watch the anarchists put on a different shirt," [former Homebrewer] Dan Sokol later recalled. "People stopped coming. Homebrew . . . was still anarchistic: people would ask you about the company, and you'd have to say, 'I can't tell you that.' I solved that the way other people did—I

didn't go. I didn't want to go and not tell people things. There would be no easy way out where you would feel good about that. ..."

It no longer was essential to go to meetings. Many of the people in companies like Apple, Processor Tech, and Cromemco were too damned busy. And the companies themselves provided the communities around which to share information. Apple was a good example. Steve Wozniak and his [friends and employees] Espinosa and Wigginton, were too busy with the young firm to keep going to Homebrew.

In the open source software world, analogous tensions arise between programmers who think a particular program should be freely modifiable and reusable, and those who would allow a business or person to have intellectual property rights over it. The subject of licensing is complicated and philosophical, but the Free Software Foundation classically defines and defends the free software concept, and private businesses take an interest in ownership of software code, and there are a spectrum of views regarding various programs.

VIII. A MODELING ENTREPRENEURIAL EXITS

In each of the historical episodes, firms burst out from networks of tinkerers to create an industry. The transition is complicated. One altered assumption can make it happen mechanically in the model. Earlier the assumption was made that the tinkerer could not see how to implement a marketable form of the technology. One might say that a veil blocks the tinkerer's view of better forms of the technology. If that veil were to lift, the tinkerer might see how to produce a product. Substantively, the new perception or belief about making an implementable product might be caused by advances in the technology, or changes externally, or by internal reflection. For simplicity, in this section the probability that the veil lifts each turn is fixed, exogenous, and known to the agent.

A generic derivation will help incorporate this into the model. Consider a one-time utility payoff which arrives with probability π at the beginning of each future period. Denote the unknown random period in which it arrives by s. We can compute the mean discount factor to apply to this payoff, $E[\beta^s]$. It is the probability-weighted average of the appropriate discount rates for each possible s. The time series summation trick is used again:

$$E[\beta^s] = \sum_{t=0}^{\infty} Pr(s == t)\beta^t$$

$$= 0 + \pi\beta + \pi(1-\pi)\beta^2 + \cdots + \pi(1-\pi)^{t-1}\beta^t + \ldots$$

$$= \pi\beta + (1-\pi)\pi\beta^2 + (1-\pi)^2\pi\beta^3 + \ldots$$

$$= \pi\beta[1 + (1-\pi)\beta + (1-\pi)^2\beta^2 + \ldots]$$

$$= \pi\beta \sum_{t=0}^{\infty} [\beta(1-\pi)]^t$$

$$= \frac{\pi\beta}{1-\beta(1-\pi)} = \frac{\pi\beta}{1-\beta+\beta\pi}$$

Suppose at some point a tinkerer (or an entrepreneur advising the tinkerer) envisions a directed research and development process which would result in a profitable product or service based on project A. Suppose further that if the tinkerer were to continue to share experimental findings universally, this would reduce the utility of the resulting monopolistic profits by more than the utility of staying in the network, so the tinkerer wishes to drop out of the tinkerer's network. Dropping out means entering a new game, in exchange for losing the payoffs a_0, ceasing to tinker with A, and ceasing to receive inflows of information from the other tinkerers, but continuing to use information from past investments and inflows.

Let M be the present utility payoff of a large monopoly profit minus the utility cost of directed research and development, capital costs, risks, and the value of the future inflows of information that would have come from the network of tinkerers, all computed at the instant the tinkerer exits the network. Let π_1 be the probability each turn that this tinkerer sees a opportunity to take M, and π_2 be the probability that the partner tinkerer does. Assume that these events cannot occur in the same period. For intuition, assume these are small probabilities.

The time-zero present value of this prospective exit in unknown period s, is M discounted by $E[\beta^s]$, which is M times $\frac{\pi\beta}{1-\beta+\beta\pi}$ as calculated above. The utility value of tinkerering up until s is

$$\frac{a_0}{1-\beta} - E[\beta^s]\frac{a_0}{1-\beta} = (1 - \frac{\pi\beta}{1-\beta+\beta\pi}) * (\frac{a_0}{1-\beta}) = (\frac{1-\beta}{1-\beta+\beta\pi}) * (\frac{a_0}{1-\beta}) = \frac{a_0}{1-\beta+\beta\pi_1}$$

Network of Tinkerers: A Model of Open-Source Technology Innovation

The mean utility cost of tinkering each period until s, falls analogously to $\frac{1}{1-\beta+\beta\pi}$. The mean benefits expected from tinkering each period until s fall to $\frac{p_1}{(1-\beta)^2} - E[\beta^s]\frac{p_1}{(1-\beta)^2} = \frac{p_1\beta}{(1-\beta)(1-\beta+\beta\pi_1)}$. The inflow of information from the partner is cut off if either partner exits, so s arrives with probability $(\pi_1 + \pi_2)$ each turn until the end. Putting that into the generic derivation, the capital value of inflows from other tinkerers falls to $\frac{fp_2\beta}{(1-\beta)(1-\beta+\beta\pi_1+\beta\pi_2)}$. Combining these pieces, the overall expected utility from joining the network is now

$$EU_0 = \frac{a_0-1}{1-\beta+\beta\pi_1} + \frac{p_1\beta}{(1-\beta)(1-\beta+\beta\pi_1)} + \frac{fp_2\beta}{(1-\beta)(1-\beta+\beta\pi_1+\beta\pi_2)} + \frac{\pi_1\beta M}{1-\beta+\beta\pi_1} \quad (4)$$

The first three terms now incorporate the possibility that these streams of utility will end, and the fourth term incorporates the new payoff of leaving the network to take payoff M.

The previous results extend forward analogously with this adjusted discounting. The net benefit of redesigning, standardizing, or specializing to raise communication efficiency to f_2 becomes $\frac{p_2\beta(f_2-f)(1-\pi_1-\pi_2)}{(1-\beta)(1-\beta+\beta\pi_1+\beta\pi_2)} - c_s$.

The net benefit of joining the network is $\frac{fp_2\beta(1-\pi_2)}{(1-\beta)(1-\beta+\beta\pi_2)} - c_j$.

For the tinkerer to prefer to exit the network when offered M, M must be at least as great as the right side of equation **4**, since at that level the tinkerer is indifferent between taking it or continuing in the network. For the story to hold together, the exit value parameter M must satisfy:

$$M \geq \frac{a_0-1}{1-\beta+\beta\pi_1} + \frac{p_1\beta}{(1-\beta)(1-\beta+\beta\pi_1)} + \frac{fp_2\beta}{(1-\beta)(1-\beta+\beta\pi_1+\beta\pi_2)} + \frac{\pi_1\beta M}{1-\beta+\beta\pi_1}$$

from which one can derive the minimum value of M:

$$M \geq \frac{a_0-1}{1-\beta} + \frac{p_1\beta}{(1-\beta)^2} + \frac{fp_2\beta(1-\beta+\beta\pi_1)}{(1-\beta)^2(1-\beta+\beta\pi_1+\beta\pi_2)}$$

Using the previous parameters $\beta = .95$, $a_0 = 1$, $f = .5$, $f_2 = .54$, and $p_1 = p_2 = .07$, here is how the payoffs change when the possibility of exits is included in a tinkerer's forecasts:

The payoffs of being in the network are thus somewhat lower if the tinkerers expect members to exit. Still, they are positive, so tinkerers would be willing to network in the near run if the entry price is low enough. Even if the tinkerers expect to be in competition with one another, the network might still hold up, depending on the parameters. To include this aspect would complicate the model and is not attempted here. It does not seem to be very important in the historical cases under consideration.

The Wright Flyer Company did not compete mainly with others who had previously been connected to Chanute.

Concept	Expression	Without exits ($\pi_1 = \pi_2 = 0$)	With exits ($\pi_1 = \pi_2 = .01$)
Utility cost of future investments	$\frac{1}{1-\beta+\beta\pi_1}$	-20	-16.81
Present value of own future progress	$\frac{\beta p_1}{(1-\beta)(1-\beta+\beta\pi_1)}$	26.6	22.35
Present value of future inflows	$\frac{\beta f p_2}{(1-\beta)(1-\beta+\beta\pi_1+\beta\pi_2)}$	13.3	9.64
Present value of standardizing	$\frac{\beta p_2(f_2-f)}{(1-\beta)(1-\beta+\beta\pi_1+\beta\pi_2)}$	1.064	.771
Minimum payoff worth exiting for	minimum M	39.9	38.07

The early Apple Computer did not compete mainly with other Homebrew Computer Club alumni. Open source software companies are in practice cooperating as well as competing with the same network their founders were in before they started their company. In these empirical cases, progress is more important than competition in the mind of the tinkerer.

There are also more differentiated outcomes in real open source software situations than the binary choice of exiting or staying in the network which was modeled. For example, the source code to the operating system Linux is freely available on the Internet, but companies such as Red Hat and SuSE/Novell develop and distribute it, and offer complementary products and services. There are a variety of licenses for open source software which keep some of the source code in the public domain. These nuanced arrangements reduce the conflict inherent in the choice as it was modeled.

The model makes explicit how tinkerers make progress *before* the industry starts, according to utility maximization, not market criteria.

IX. Relaxing the Assumptions

IX.A. Rates of Progress

The assumption that each tinkerer achieves a high, steady, known rate of progress can be relaxed in some cases, and still allow a tinkerer's network to hold.

First, the assumption that progress occurs at a known fixed rate is stronger than necessary, although it is a useful simplification. A more realistic description is that tinkerers see some stream of opportunities to achieve progress as they define it. They have informed expectations about experiments, based on their knowledge and experience. A tinkerer tries experiments, whose outcomes are random. Tinkerers quit if dissatisfied with their progress. By self-selection, the population of tinkerers tends to consist of those who can make effective progress, and the p in the model is a long-run average for each member of this selected population. Modeled in this way, the present value of utility could not be so readily computed analytically. The complexity thereby introduced would distract from the main points of this model.

Second, the assumption that each tinkerer achieves a high enough rate of progress alone to motivate his own efforts is not always necessary. Tinkerers could play other roles in a network. Here are two examples:

- Suppose a tinkerer is in two networks which address different kinds of projects but occasionally some idea in each one is useful to tinkerers in the other. As an information broker transmitting these cross-cutting ideas, a tinkerer may contribute enough to maintain membership in both networks.
- A family member or friend of a tinkerer may encourage a tinkerer, express interest in the project, and bounce ideas back and forth with the tinkerer. This helpful person would not need to make any specific rate of progress or pay any joining cost to in essence become part of the network, learn about the project, and seize on useful opportunities to help the tinkerer. the progress that the tinkerer makes. Among the aerial experimenters there were several pairs of brothers among the aerial experimenters. At least in the case of the Wrights, the close collaboration helped the Wrights stick with the project through bad

patches in which there was not much success. Thus other relationships can support a network's relationships.

Thus the assumption that every individual's progress outpaces a discount rate is not strictly necessary. The essential assumptions are that tinkerers are interested in common projects and can make mutually helpful progress on them according to their own judgement.

IX.B. Technological Uncertainty

The model assumes an agent cannot sell the technology at a profit, and cannot foresee how such possibilities could occur in the future. This is an extreme version of *technological uncertainty*, described in Tushman and Anderson (1986), Dosi (1988), and Rosenberg (1996). If there is no technological uncertainty, and the path to a marketable design is clear, then, by perfect foresight arguments, a profit-seeking firm would do that immediately. So if tinkerers lead the way technologically to a profitable industry, there must have been technological uncertainty.

The model assumed that tinkerers operate under technological uncertainty and so could not see how to make a version of A for which there would be enough demand to make a profit. In casual conversation one might say that he does not see a version of A that is "good enough" to sell, but with radically new products, both supply and demand may be hard to foresee. Several early aircraft developers did not expect the rapid military adoption of aircraft. Early personal computer makers dramatically underestimated demand. Tushman and Anderson (1986) used errors in forecasts by industry analysts as a metric of technological uncertainty.

Because of this, investment and payback for tinkerers are unavoidably subjective in this model. The experimenters do not know future forms of the output (whereas we can look *back* at a well-defined "invention of the airplane"). The improvements include qualitative redesign and "failed" experiments. A tinkerer may expect to have a better understanding of the activity after an experiment, whether or not it improves A in functional terms, and it may benefit other tinkerers to know about that experiment. Therefore the model incorporates subjective progress, and does not measure progress by engineering or market attributes of A.

IX.C. Scale and Population Size

The model has only one relationship between two tinkerers, but each can be connected to groups or networks beyond. So the network model can scale up. As modeled, all participants contribute information to create a positive sum interchange, potentially having positive externalities. There is positive feedback, because fast progress makes a network more appealing to join. Its expansion is limited because experimenters in frontier technologies are rare.

For tinkerers of a given level of interest and capability, a larger network makes faster progress than a small one. So members have an incentive to reduce barriers to communication within the network, or with people who might join the network. This benefit of scale implies that networks with fewer barriers will address technical problems better or more quickly than networks with more barriers. For example, the use of English to discuss an open source project may improve the speed of development if more potential programmers can participate in English. This suggests that, holding other things constant, open source innovation will tend to be more successful when tinkerers communicate in a language many people know, and in locations with less restriction on printing or association with other people. If there are many tinkerers, the network will probably have greater internal friction, and require administrative structures of information sharing (as in open source software), which are not modeled here.

IX.D. Motivation

The tinkerer is imagined to have intrinsic or altruistic motivations. But a network can form in support of a profit-making or career effort of the tinkerer too, as long as the other tinkerers do not see it as a zero-sum competition. So for example a government laboratory's programmers may contribute to an open source software development effort if it is useful to their project, or a cost-reducing effort could be co-sponsored by competing firms.

A tinkerer's motivation could include not only the possible honor of making a major invention, but also the possible second-best prize of being recognized and cited by the final inventor. Such streams of payoffs can be viewed as a portion of a rate of progress p. The model also excludes any payoffs experienced in sharing what one has done with others, although innovators report the opportunity to share is beneficial and satisfying.

In the examples that motivate this model, the tinkerers do not yet know much about how to make a good A because the technology cannot yet be usefully implemented. The number of tinkerers who can make experimental progress on a particular type of project is limited to those with the knowledge, wealth, and tools to attempt it. Many people could value experimenting with new aircraft, but few like it enough, and are good enough at it, and have the resources, to bother. Those few have opportunities to make something that looks like progress to them. One might imagine that values of a_0, the original payoff of activity A, were drawn from a distribution, and the few people with a sufficiently positive value for a_0 would be tinkerers. In the aircraft case, even successful experimenters considered quitting, and many did.

Once a technology is established and competitively produced, the basic uncertainties have been resolved, so the model loses relevance. Today, there is an established spectrum of technologies for making aircraft and personal computers and delivering services to and from them. Technological uncertainty still exists within narrower domains that could be relevant for the model.

IX.E. Enforceability

In the model, tinkerers would be willing to agree to an enforced open-source contract rather than work alone. But in the case of gliders leading to the airplane, there was no exogenous enforcement. There are several sources of incentive besides technological progress to support an open-source pattern which were not modeled here.

- Innovators may feel an obligation to share with the group, so enforcement is internal to each tinkerer. Meyer (2003) discusses some examples.
- Innovators may want their peers to see their work because they are proud of it and will be favorably recognized for it, as discussed in Raymond (2001) and Levy (2001). Unlike a_0 in the model, this payoff directly supports open-source relationships, and does not depend on an information flow back from the other person.
- If an invention delivers more output when it has more users, inventors may benefit more by giving it away than by keeping it secret or charging a fee to examine or use it. For example, since Web browsers

were invented, many have been given away to make collaboration and information tracking easier (Berners-Lee, 1999).
- In the model of Bessen and Maskin (2006), profit-making firms are willing to share innovation information openly with one another if they are following different paths of research or if the innovations they expect to make will be useful to achieving future ones. For tinkerers, one might model this by raising the rate of progress each tinkerer expects if more sources of information are available.
- Tinkerers may gain more from interaction if they are familiar with one another's work. Adapting this model, f could rise over time as the network's members develop longer histories together.
- To a tinkerer who anticipates someday selling a product or service, the population of tinkerers inside the network may be the natural market of early adopters for it. Interaction with others helps the tinkerer know what customers will want, and creates an opportunity to earn their trust. At first only specialists could understand and appreciate the aircraft, personal computers, and new types of software discussed earlier.

Given incentives like these, sustaining the agreement can be rational for each individual participant. The enforceable contract in this paper is a modeling shortcut. More accurate stories may require a more complex model.

IX.F. Frictions, Diseconomies of Scale, or Other Costs

As written, the model incorporates the extreme assumption that there are no economies or diseconomies of sharing with more and more people. For example, it is implicitly assumed that there is no time constraint in keeping up with the relevant literature, nor cost for communicating to yet another person, nor changing marginal cost to enforce the sharing agreement. One could incorporate such influences by making f a function of the number of participants to generate increasing returns to scale (encouraging evangelism) or decreasing returns to scale (inducing pressure to reduce or exclude members).

Many aspects of the environment affect f. If, for example, communication between the tinkerers are noisy or clogged with unhelpful communication, such as email spam, f is lower. If the languages of technical communications are different, f is lower. An American experimenter working on gliders may naturally choose not to read a French journal about balloon developments, even if the balloon work is productive in its own terms (measured by p_2), either because he cannot read French, or because he thinks balloon innovations are unlikely to apply to gliders.

The model implies there are no expensive capital or training requirements. In the examples, tinkering was not usually capital-intensive. It appears that once expensive equipment is necessary for some activity, that activity falls beyond the boundaries of a network of tinkerers, except perhaps inside an organization. Also, tinkering tends to arise in environments where not much specialized training is required.

X. CONCLUSION

A network of tinkerers model applies to innovative processes in which:

- Individuals communicate novel technical findings and designs about a technology to one another without explicit compensation.
- Experimenters do not have extrinsic rewards, largely because they are working on something that has no obvious price or does not fit into an existing, standard product market at the time they enter the field.
- Some participants specialize in managing or expanding the network.
- The activity evolves over time, in response to events that participants interpret as *progress*, such as discoveries or inventions. For example, when Hargrave reported results from his box kite experiments, other aeronautical experimenters learned and adapted to the findings without imitating his experiments. Thus they behaved as if they were responding to a discovery of a natural law or invention, not performing for others or engaging in a sport.

Given such a situation, the model predicts that participants specialize in aspects of the technology, and standardize on some tools, as opportunities permit. The framework assumes predictions about the future form or importance of the technology are diverse and uncertain in the sense of Dosi (1988) and Rosenberg (1996). It predicts that members who do not plan to sell a related product quickly will object to intellectual property impositions. And it predicts the kind of ferment which can lead participants to jump out into entrepreneurial ventures, whose value is difficult to predict in the sense used by Tushman and Anderson (1986). The model predicts progress would be slowed by high costs to invest in experimentation or to find and join networks.

Examples of this process occurred in the invention of the airplane, the invention of the personal computer, and in the development of open source softare programs. There is a spectrum of similar kinds of innovative development:

- Shared creative content can be like open-source software. The Wikipedia, for example, is a public domain encyclopedia written principally by unpaid users. The collection of video content at YouTube.com is donated by users. In these cases the pooled content is not made up of functional engineering achievements. Instead the developers are sharing text, reasoning, and media. The library grows with contributions from many users to advance in a direction they more or less agree upon. Such shared content is similar to open source software.
- **Open science.** The rise of "open science" institutions which motivate, support, and enforce open publication of scientific findings, was supported by competition between patrons to employ prestigious and effective scientific innovators (David, 1998). Through the network of tinkerers story, open science would also be supported by people who wish to speed the progress of science.
- User innovation, in which a company produces a product and its users generate *user innovations*, as defined and discussed in von Hippel (2006), which also provides many examples.
- **"Skunkworks", projects of creative engineering** inside corporations in which engineers work around an employer's hierarchy, rather than obeying it. Their goal may be for the organization to succeed despite its managers. The model offers a way to think of such actors less as shirking, and more like visionaries, as they think of themselves.
- The British Industrial Revolution after 1750 occurred in a place and time when printers were allowed more freedom about what to print than printers elsewhere. This helped support an estimated 1020 technical and scientific societies (Inkster, 1991, pp 71-79). Workshops of the time were often open to visitors, and they generated, by one account, **"a wave of gadgets [which] swept [over] Britain"** (Mokyr, 1993, p. 16, citing Ashton). Mokyr (1993, p. 33) concluded: "The key to British technological success was that it had a comparative advantage **in microinventions.**"

- The Internet and the Web expand the opportunities for technological discussion.

This model provides a formal structure to describe developers of radically new technology. Some innovators make fast progress on their own, although it may not look like progress to other people. Others make sufficient progress to perservere, if they have exogenous links to one another. Independent innovators have an incentive to join networks and share flows of information. An innovator with comparative advantages in recruitment, publicity, moderating conversation, publishing, or editing journals may end up doing those things because that moves the project forward faster than if this person worked directly on the technology. Much of this can be understood and discussed in terms of progress rates p, fractional flows of useful information f, and differentiated opportunities for each person. Players may imagine profitable future exit opportunities but if these are remote or improbable enough, joining the network makes sense until participants actually see those opportunities.

Tinkerers in the model choose to combine their information to maximize the combined flow of useful innovations. The speedup in the flow of innovations is therefore endogenous. In the model, purposeful choice generates flows of innovation that other economic models of technological change often take as given. When technological uncertainty is great, tinkerers networks can eventually form an industry, once enough tinkerers can anticipate commercial possibilities from their activities.

The desire of people to make their world a better place is a kind of natural resource. The environment affects their effectiveness. If publishing a journal, forming an association, and traveling are costly or officially discouraged, innovators facing technological uncertainty would be less effective. In this model that would reduce their utility. In the real world they might respond by reducing effort, keeping innovations secret, or emigrating to a location where the environment was more favorable. So noisy or restricted communications channels can reduce the flow of innovations both by reducing the flow of communication, and by driving tinkerers away.

In this model, innovation is generated by individuals not organizations. One benefit of modeling innovation this way is that the predictions and intuitions often apply outside the context of businesses and hierarchies, as in production prior to capitalism, or communications inside or between organizations.

XI. REFERENCES

Allen, R. C., (1983). Collective invention. *Journal of Economic Behavior and Organization 4*, 1-24.

Bessen, James & Eric Maskin. (2007). Sequential Innovation, Patents, and Imitation. Institute of Advanced Study School of Social Science Working Paper 25. 2007. Forthcoming *RAND Journal of Economics*.

Berners-Lee, Tim. (1999). *Weaving the Web*. HarperCollins.

Chanute, Octave. (1997). *Progress in Flying Machines*. Dover, Toronto.

Cringely, R. X. (1992). *Accidental Empires*. HarperCollins edition, 1996.

Crouch, Tom. (2002). *The Dream of Wings: Americans and the airplane, 1875-1905*, second edition. Norton.

David, Paul A. (1998). Common Agency Contracting and the Emergence of "Open Science" Institutions. *American Economic Review*, May 1998, 15-22.

Dosi, Giovanni. (1988). Sources, Procedures, and Microeconomic Effects of Innovation. *Journal of Economic Literature 26:3* (Sept.), 1120-1171.

Freiberger, Paul & Michael Swaine. (1984). *Fire in the Valley: the making of the personal computer*, first edition. McGraw-Hill.

Harhoff, Dietmar, Joachim Henkel, and Eric von Hippel. (2003). Profiting from voluntary information spillovers: How users benefit from freely revealing their innovations, *Research Policy vol 32*, No.10 (Dec), 1753-1769.

Inkster, Ian. (1991). *Science and Technology in History*. Rutgers University Press.

Jakab, Peter. (1990). *Visions of a Flying Machine*. Smithsonian Institution.

Lakhani, Karim R. & Bob Wolf. (2005). Why hackers do what they do: Understanding motivation and effort in free/open source software projects. In *Perspectives on Free and Open Source Software*, edited by J. Feller, B. Fitzgerald, S. Hissam, and K. R. Lakhani, 2005, MIT Press. Available at http://opensource.mit.edu/papers/lakhaniwolf.pdf

Levy, Stephen. (2001). *Hackers: Heroes of the Computer Revolution*. Penguin edition.

Meyer, Peter B. (2003). Episodes of collective invention. BLS Working paper WP-368. Available on Web at http://www.bls.gov/ore/abstract/ec/ec030050.htm.

Mokyr, Joel. (1990). *The Lever of Riches: technological creativity and economic progress*. Oxford University Press.

Mokyr, Joel. (1993). Editor's introduction. *The British Industrial Revolution: an economic perspective*. Westview Press.

Nuvolari, Alessandro. (2002). Open Source Software Development: Some Historical Perspectives. ECIS working paper.

Pavlicek, Russell C. (2000). *Embracing Insanity: Open Source Software Development.* Sams Publishing.

Raymond, Eric S. (2001). *The Cathedral & the Bazaar: Musings on linux and open source by an accidental revolutionary.* O'Reilly.

Rosenberg, Nathan. (1996). Uncertainty and technological change. *In Mosaic of Economic Growth, edited by Ralph Landau, Timothy Taylor, and Gavin Wright.* Stanford University Press.

Schrader, Stephan. (1991). Informal technology transfer between firms: cooperation through information trading. *Research Policy 20:*153-170.

Tushman, Michael L. & Philip Anderson. (1986). Technological Discontinuities and Organizational Environments. *Administrative Science Quarterly* 31 (Sept.): 439-465

von Hippel, Eric. (2005). *Democratizing Innovation.* MIT Press.

Wozniak, Steve, with Gina Smith. (2006). *iWoz*, Norton and Company, New York.

Wright, Orville. (1953). *How We Invented the Airplane.*

Wright, Wilbur. (1901). Angle of Incidence. *The Published Papers of Orville and Wilbur Wright*, edited by Peter L. Jakab and Rick Young, 2000.

CHAPTER SOURCES

The following chapters have been previously published:

Chapter 1 – This is an edited, excerpted and augmented edition of a Small Business Administration, Office of Advocacy Report, under contract no. SBAHQ07-Q-0012, dated March 2009.

Chapter 2 – This is an edited, excerpted and augmented edition of a United States Department of Labor, U.S. Bureau of Labor Statistics, Office of Productivity and Technology, Working Paper 413, dated November 2007.

INDEX

A

accounting, 5, 13
administrative, 53
aeronautical, 33, 56
agents, ix, x, 31, 32, 41
air, 36, 42
Airplanes, x, 31
alternative, 10, 38
ambiguity, 11
application, 3, 13, 26, 27
assets, 13, 24
assumptions, 32, 40, 41, 52
asymptotic, 12
autocorrelation, 10, 13, 29
availability, 3

B

back, 2, 8, 51, 52, 54
balance sheet, 13
banks, 22, 24, 25
barriers, 53
BEA, 31
behavior, 11, 45
benefits, 14, 35, 36, 38, 40, 41, 42, 43, 45, 49
biotechnology, 2, 3
blocks, 47
Britain, 33, 57

brothers, 46, 51

C

candidates, 44
capital cost, 48
capitalism, 58
causality, 29
channels, 58
chemicals, 20
classes, 20
clusters, 29
collaboration, 51, 55
combined effect, 9
commercialization, 32, 45
communication, 42, 49, 53, 55, 58
communications channel, 58
community, 37, 46
comparative advantage, 57, 58
compensation, 56
competition, 16, 43, 45, 46, 50, 53, 57
complementary products, 50
complexity, 51
components, 21, 37
compounds, 20
computer systems, 34
computer technology, 34
computing, 34
conflict, 50
consensus, 43

construction, 18, 21, 37
consumption, 20
control, 9, 11, 35
copyrights, 35
corporations, 57
correlation, 11, 16
cosmetics, 20
costs, 23, 25, 38, 40, 41, 42, 43, 44, 45, 48, 57
creativity, 59
credit, 22, 23
cross-sectional, 10
customers, 23, 55

D

data gathering, 6
data processing, 22
database, 3, 6, 7, 8, 29
decisions, 38
decreasing returns, 55
definition, 42, 45
dependent variable, 9, 10, 12, 13
depreciation, 29
discount rate, 47, 52
discounting, 49
distribution, 11, 12, 21, 54

E

earnings, 16, 24, 29
economic development, 28
economic growth, 19
Education, 27
electricity, 21
empirical studies, 5
employee headcount, ix, 1, 2, 4, 8, 9, 10, 11, 12, 15, 29
employees, 2, 4, 6, 7, 8, 9, 16, 22, 23, 46, 47
employment, 6, 9, 19, 26
employment growth, 19
energy, 21
engines, 18, 21
enthusiasm, 36
environment, 44, 55, 58

excise tax, 24
expenditures, ix, 1, 2, 3, 4, 6, 8, 11, 12, 13, 14, 15, 18, 19, 20
explosives, 20
extrinsic motivation, 36
extrinsic rewards, 56

F

failure, 44
family, 51
feedback, 53
fees, 24
fertilizers, 20
fibers, 20
finance, 24
financial performance, 4, 5
financing, 10
firm value, 1, 2, 3, 4, 19
firms, 2, 3, 4, 5, 6, 10, 12, 15, 16, 18, 19, 22, 32, 36, 47, 53, 55, 60
fixed rate, 51
flow, 9, 34, 38, 41, 42, 43, 45, 46, 54, 58
focusing, 2, 19
France, 33
franchise, 24
freedom, 57
friction, 53

G

generally accepted accounting principles, 13
geophysical, 22
Germany, 33
government, 3, 5, 19, 25, 53
grants, 2, 3, 6
Gross Domestic Product, 4
groups, 32, 53
growth, ix, 1, 5, 16, 19, 27
guidance, 22

H

hackers, 36, 37, 59

Harvard, 26, 28
heteroscedasticity, 10, 13, 29
household, 21
human, 35, 36
human condition, 36
humanity, 36
hydrological, 22

I

identification, 7
incentive, 42, 53, 54, 58
incentives, x, 31, 45, 55
incidence, 42
income, 23, 24
increasing returns, 55
independent variable, 9, 10, 11, 15
indicators, 5
industrial, 18, 21
industry, ix, x, 2, 3, 4, 13, 14, 16, 17, 18, 19, 21, 23, 25, 29, 32, 34, 47, 50, 52, 58
inflation, 29
information sharing, 53
innovation, ix, 1, 2, 3, 4, 5, 6, 8, 9, 10, 11, 12, 16, 20, 26, 29, 31, 40, 41, 53, 55, 57, 58
institutions, 35, 45, 57
instruments, 22
insurance, 24
intangible, 11, 12, 13
intellectual property, 8, 45, 47, 56
Internet, 34, 44, 50, 58
intrinsic, 36, 53
intrinsic motivation, 36
inventions, 56
investment, 20, 38, 39, 40, 43, 52
investors, 14
isolation, 44

J

jobs, 5
joining, 41, 43, 49, 51, 58

L

labor, 3
language, 42, 53
learning, 9, 46
license fee, 24
licenses, 50
licensing, 8, 44, 47
linear, 12, 14, 45
linear function, 12
linear model, 12, 14
links, 58
Linux, 36, 50
liquor, 24
location, 58
losses, 24, 42

M

machinery, 18, 21
machines, 33
magazines, 34
management, 12
manufacturing, 20, 21, 22, 46
market, 1, 2, 3, 4, 5, 6, 8, 11, 12, 13, 16, 19, 29, 38, 43, 45, 46, 50, 52, 55, 56
market value, 2, 3, 6, 8, 11, 12, 13, 16, 19, 29, 38
marketplace, 14
Maximum Likelihood, 27
measurement, 11, 12
measures, 4, 11, 20
membership, 34, 51
meteorological, 22
metric, 45, 52
military, 52
mining, 18, 20, 21, 25
MIT, 28, 59, 60
modeling, 55, 58
models, 9, 26, 27, 29, 58
moderators, 44
monopoly, 48
Monte Carlo, 26
motivation, 36, 37, 53, 59
motors, 21

movement, 36
multivariate, 2

N

natural, 13, 14, 20, 43, 44, 55, 56, 58
network, ix, x, 31, 32, 35, 36, 40, 41, 43, 44, 45, 48, 49, 50, 51, 53, 55, 56, 57, 58
networking, 45
New York, 27, 60
newsletters, 34
niche market, 5

O

obligation, 54
observations, 4, 11
oil, 18, 21, 23, 24
ophthalmic, 22
opportunity costs, 38
optical, 22
organic, 20
organic chemicals, 20
originality, 5
outliers, 14
ownership, 47

P

packaging, 20
paints, 20
parameter, 12, 13, 38, 44, 49
Pareto, 44
Patent and Trademark Office, 3, 6
patent policy, 5
patents, ix, 1, 2, 5, 7, 8, 11, 13, 15, 16, 17, 27, 45
pay off, 45
perception, 47
periodic, 38
permit, 56
personal computers, x, 31, 54, 55
philosophical, 47
pigments, 20
plastics, 20
play, 4, 5, 51
population, 45, 51, 55
portfolios, 5
positive externalities, 53
positive feedback, 53
positive relation, 4, 5, 15, 19
positive relationship, 4, 5, 15, 19
potassium, 20
powder, 20
premium, 11, 17
present value, 40, 48, 51
pressure, 55
prestige, 36
private, 4, 19, 29, 45, 47
private ownership, 45
private sector, 4, 19
probability, 12, 47, 48, 49
probability distribution, 12
product market, 56
production, 2, 3, 4, 9, 11, 12, 13, 14, 15, 16, 17, 29, 58
productivity, ix, 27, 32
productivity growth, 27
profit, ix, x, 24, 29, 31, 32, 34, 40, 48, 52, 53, 55
program, 35, 38, 47
programming, 22
property, vi, 8, 11, 12, 24, 35, 45, 47, 56
property rights, 45
proposition, 40
protection, 2, 3, 17, 45
proxy, 2, 3, 4, 9, 11, 12, 13
public, ix, 6, 8, 29, 36, 42, 50, 57
public companies, 6, 8
public domain, 36, 50, 57

R

R&D, ix, 1, 2, 3, 4, 6, 7, 8, 11, 12, 13, 14, 15, 18, 19, 20, 25, 26, 27, 28
random, 47, 51
reality, 33
reasoning, 57
reflection, 47

Index

refrigeration, 21
regression, 2, 9, 10, 11, 12, 14, 15
regression analysis, 2
regular, 23
relationship, 2, 3, 4, 5, 8, 9, 12, 14, 15, 16, 19, 20, 27, 29, 53
relationships, 8, 9, 52, 54
repackaging, 20
reproduction, 22
research and development, ix, x, 1, 29, 32, 48
Research and Development, 18, 25, 28
resources, 3, 54
retained earnings, 25
revenue, 23, 24, 29, 36
rewards, 5, 37
royalties, 25
rubber, 24

S

sales, ix, 1, 2, 4, 6, 8, 9, 10, 12, 14, 15, 16, 18, 19, 23, 24, 29
salt, 20
salts, 20
sample, 5, 8, 16
satisfaction, 38
schema, 37
search, 7, 22, 43, 44, 46
search engine, 7
secrets, 13, 46
securities, 24
sensors, 22
series, 10, 29, 34, 38, 47
services, 1, 22, 25, 50, 54
shareholders, 1, 22, 26, 46
shares, 26, 29
sharing, 32, 33, 34, 35, 43, 44, 45, 46, 53, 55, 57
shocks, 11
signals, 12
silicon, 20
skilled labor, 3
Small Business Administration, 26, 28, 29, 61

small firms, 3, 4, 5, 15, 16
Smithsonian, 37, 59
social benefits, 45
sodium, 20
software, x, 25, 31, 32, 35, 36, 37, 42, 43, 44, 47, 50, 53, 55, 57, 59
software code, 47
specialization, 43
spectrum, 35, 47, 54, 57
speed, 44, 53, 57
spillovers, 3, 6, 27, 59
Standard error, 17, 18
standardization, 42, 43
startup firm, x, 32
statistics, 19
stock, 4, 6, 25, 29
stock price, 29
storage, 21
strategies, 16
streams, 49, 53
subjective, 40, 52
subsidies, 24
substitutes, 43
supply, 22, 52
surgical, 22
surplus, 25, 39
survival, 2
survival rate, 2
synthetic fiber, 20

T

talent, 37
taxes, 24
technological change, 5, 31, 58, 60
technological progress, 54
technology transfer, 60
television, 21
territory, 34
thinking, 42
time frame, 5
time periods, 10
time series, 38, 47

tinkerers, vi, ix, x, 31, 32, 37, 40, 41, 42, 44, 45, 47, 48, 49, 50, 51, 52, 53, 54, 55, 56, 57, 58
tracking, 55
trade, 13, 23
trading, 60
training, 56
transactions, 8
transfer, 60
transformation, 21
transition, 47
transmission, 21, 45
transmits, 45
trucks, 18, 21

U

U.S. economy, 4, 6, 19
uncertainty, 52, 54, 58
unemployment, 16
United Kingdom, 27
United States, 11, 28, 33, 61

V

valuation, 2, 3, 11, 44
values, 12, 18, 19, 25, 38, 43, 54
variables, 8, 9, 10, 11, 29
variance, 29
variation, 16

W

watches, 22
wealth, 1, 36, 54
wind, 33, 42, 46
winter, 46
workers, 22, 23
World Wide Web, 36
writing, 38, 44

X

XRD, 25